A Phenomena-Based Physics

Sound ~ Light ~ Heat

Volume 3 of 3
for

Grade 8

Manfred von Mackensen

Waldorf
PUBLICATIONS
RESEARCH INSTITUTE FOR WALDORF EDUCATION

Sound ~ Light ~ Heat

Electricity, Magnetism & Electromagnetism,
Mechanics, Hydraulics and Aeromechanics
as Subjects in an
Introductory Physics
6th through 8th Grades

Volume 3 of 3
for

Grade 8

by
Manfred von Mackensen
Pedagogical Research Department
Union of German Waldorf Schools
Kassel, West Germany, 1992

translated and edited by John H. Petering
August 1994

Printed with support from the Waldorf Curriculum Fund

Published by:
Waldorf Publications at the
Research Institute for Waldorf Education
351 Fairview Avenue, Suite 625
Hudson, NY 12534

Title: *A Phenomena-Based Physics: Sound, Light, Heat: Grade 8*
Author: Manfred von Mackensen
Editors: David Mitchell and John Petering
Translation: John Petering

German original copyright 1987, 1992 by Manfred von Mackensen,
Kassel, West Germany
First printing © 1994 by AWSNA Publications

Revised edition © 2019
ISBN #978-1-943582-24-2
Layout: Ann Erwin
Cover: Postcard picturing the Peterboro hydraulic lift lock, Ontario, Canada,
the largest in the world

Table of Contents

Contents

GRADE 8 ELECTRICITY

GRADE 8 HYDRAULICS/HYDROSTATICS

GRADE 8 AEROMECHANICS

GRADE 8 ACOUSTICS

ENDNOTES

SELECTED BIBLIOGRAPHY

Introduction

I. PREFACE

The following chapters describe a selected area of teaching—physics—the contents, initial thoughts, yes even the entire construction of the teaching material and many teaching methods, all drawn from the treasure chest of Waldorf pedagogy. But, simply by working through the descriptions for an *individual* physics topic given here, someone foreign to the Waldorf methodology can hardly achieve intimacy with the whole of this pedagogy. And yet, it is only from this whole that such particulars (e.g., the physics lessons) can arise or be available.

What is presented here can be only a kind of *indication* of this whole, that at first stands open and unknown. At the same time, this material *can* serve as a stimulus also for the teaching efforts of those not familiar with Waldorf education. The present material could even be useful from a variety of viewpoints, e.g., finding little-known physics experiments.

Still, in spite of these obstacles, the fundamental concepts of the Waldorf approach to natural science should reach a wider audience. Certainly, it will be in the spirit of the *preconceptions of the author* and the working-group here in Kassel (the basis for this whole project), including their understanding about scientific theories, developmental psychology and commercial starting points—insofar as an introduction can draw comparisons to contemporary ideas.

Although this book alone is not really an introduction to the essentials of Waldorf pedagogy, nor can it even fully develop a scientific foundation for our approach, yet we must attempt to highlight our preliminary understanding, which has been used effectively, and draw attention to it in the context of the contemporary science education literature. The starting points of our work should be made visible; thereby, skeptical readers will be able to select something useful, try it out, and perhaps have it make sense sooner than later.

Whoever is new to this subject, or initially wants to spare themselves a discussion of foundational scientific theory and educational psychology (possibly many readers), could perhaps read only the final section of this introductory chapter: "VII. To the Teacher," which summarizes many theoretical and practical points. After that, they could experiment with the various lesson outlines of a phenomena-based approach to physics. Thus, they will have the necessary examples in front of them, and later may go back to read the rest of the background to this approach to studying the sciences, especially physics.

[Note: The following sections have been summarized from von Mackensen's detailed discussion, since most of the educational and philosophical authors cited are not well-known in the English-speaking countries. For a full critique of the philosophical underpinnings presented by Dr. Mackensen, a full translation of his introduction is available from the Waldorf Publications clearinghouse upon request, at no cost.]

II. TEACHING, TECHNOLOGY & LIFE

Basically, we have less and less success guiding young people in learning and work, both at school and at home. Education slips away from this generation; although the amount of scientific work given to them is greater than ever before! Elaborate reforms, allocation of financial resources and dedicated administration in the schools have not helped the whole. Plainly, neither the analytical, technical, nor legal modes of thinking truly relate to those deeper aspects of the human being capable of transformation and learning. The human forces which allow people to form their spirituality, enliven their soul, and enable them to grow, plainly lie in another realm from the modern scientific way of thinking.

Educational writers themselves point to a need for a reorientation of science teaching to make it relevant to people's daily life, in view of the "lack of impact of a technically-oriented science teaching, which tries to tackle the teaching and learning process using the [quantitative] methods invented by that same science...." (Duit 1981) The fact that schools aren't easily reformable with technocratic means is due to the fact that "the first phase of reform was ... too foreign from the children, went off into abstract scientific criteria. ... Along with a structurally-rooted, growing unemployment, the principle of competition shifts down into ever lower grades. And, as schools allow such selective and competitive modes to creep in, it only increases the impoverishment of students in terms of social experiences, emotional connections, and feeling-filled learning possibilities.

"On the heels of the discovery of possibilities for early learning came performance-based teaching programs, minutely planned class periods (with well-planned teaching behavior, and expected student behavior), and increased *demands for more content* and more technical-professional scientific topics in the lessons. To our dismay today, we actually *establish the isolation and alienation of the human being.*" (Silkenbeumer 1981) Perhaps we can explore further this contrast between the bright well-defined world of science with the dark, less consciously understood world of life, albeit in a living manner, using open-ended characterizations, and always mindful of the whole range of human activity.

Another description of how these two principles from the world of life and of science infiltrate everything is given in the last great letters of Erich Fromm. He characterizes them as "having" and "being": "Indeed, to one for whom having is the main form of relatedness to the world, ideas that can not easily be pinned down (penned down) are frightening—like everything else that grows, changes, and thus is not controllable." (Fromm 1976) After showing how the current crises of humanity spring from the ideology of fixation and having, Fromm describes how the modern technological world promotes a marketing-character, people who "experience themselves as a commodity ... whose success depends largely on how well they sell themselves on the market" and, dedicated to pursuing the quantitative side of life, become *estranged* from their living surroundings and from their own self. "Most striking, at first glance, is that Man has made himself into a god because he has acquired the technical capacity for a second creation of the world, replacing the first creation by the God of traditional religion. ... Human beings, in the state of their greatest real impotence, in connection with science and technology *imagine* themselves to be *omnipotent*." (Fromm 1976a)

If we shy away from Fromm's pedagogical indications, we can still value his observations and views and note again that *it matters for the whole of humanity, for everyone.* So, whether it admits it or not, education also is tensioned between these poles of "having" and "being," or between convergent and divergent issues. (Schumacher 1979)

Convergent issues are of a type which *can* be dealt with and converge on a stable solution simply by applying more time, intelligence, or resources (even without higher forces of life, a wider consciousness, or personal experience). Dealing with *divergent* problems with more study only produces more polarity— they require a higher level of being, beyond mere logic, where the seeming opposition of polarities can be transcended. A classic example is finding the right educational methods, where the right formula or answer is seemingly never found in the opposing ideas of more form vs. more autonomy. Divergent issues involve a higher realm, including both freedom and direct inner experience.

We all must cope with our scientific world; its development is necessary for a free consciousness. But now it challenges us to find its limits so that the living environment, society, and the human spirit can breathe. **This problem is truly the deepest, most global problem of our time.** Within science, challenges lead us to new *scientific theories,* and many, diverse initiatives exist, but with only meager results up to now. (We will indicate our relation to each of these initiatives below.) When our thinking within the living world is challenged, it can lead to *Goetheanism*, to a phenomenology rich in living ideas. But, mere concerns in the living world lead only to mythology.

III. THE LANDSCAPE OF SCIENCE

PRESUMED CERTAINTY VS. LIFE

In recent years there have been many articles published that discuss the philosophy of science (How do we know the world?). Science teachers adopt the same well-accepted, positivistic habits of thought as more technical writers, presuming that conventional quantitative methods do give us a real knowledge of nature. Yet, if we delve more deeply into what many writers in the philosophy of science say about this, what emerges is not at all so cut and dried—nor clear. The fundamental question is: What is the relationship between our selves and the world out there? This is one of *the* fundamental questions of philosophy. Many philosophers such as Descartes, Kant and Goethe have wrestled with the question. Does my mind (soul, spirit) have any real connection with my body (the world of matter)?

Even contemporary educators acknowledge the subtleties of this relationship: "What scientific research achieves, in the best case, is not a plain portrayal of reality. The theory of science has shown that the researcher is not only interwoven with his experiments and the measuring apparatus he uses, but also (in a much more subtle but potent way) is intermeshed with his whole theoretical and mental framework, with his very concepts and way of questioning, with his definitions and hypotheses."[1] (Wieland 1981) And in 20th century physics itself, exact results have come about which do not illuminate a clear reality, and indeed never can. The Copenhagen Interpretation about quantum theory (1926) early on led a leading physicist to say "... the reality is different, it depends on whether we observe or not ... and we must remember that what we observe is not Nature itself, but nature *as expressed to our manner of questioning*." (Heisenberg 1959; see also K.F. von Weizsäcker 1971)

This awareness, beginning first in physics, now spreading to many other fields, forces us to recognize that our relation to the world is not as simple as we might have thought. Although there are myriad ways of considering nature, for *our* path the image of nature we develop fully develops and ripens only *through the human being*—the human being is not superfluous. Our preconceptions (what T. Kuhn calls our paradigm or framework of what we take "world" to be) matter a lot—they condition the very character of what we come up with. If we really think about it, the very reality which science presumes cannot be thought about as something real "out there," with our selves as something separate, a kind of onlooker; rather its reality consists in our mutual relationship and intimate connections with the world.

THE SOLE AUTHORITY?

While it's true that the conventional literature doesn't claim to be the only valid method, at least not *explicitly,* still, this attitude is *implemented,* in that empirical or positivistic habits are accepted and implicit in the way science and teaching are pursued. Actually, some educators have come very close to touching on these questions, and their ideas are significant not because they pose a particular problem, but because they discuss *standards* of what is "good" science.

They usually presume that our knowledge and theories relate to a pre-existing objectively material world. And "what gives [this method] certainty is a clearly marked path back to the evidence of data." (Jung 1982) This presumes an aspect of phenomena separate from our conceptualizing, which is part of an objective world, existing *before* we think about the phenomena. So "reality" must consist in a *collection of basic phenomena* (i.e., all our perceptions in the living world) which is accessible to all persons.

So the fundamental question becomes: Is there some sort of pre-existing objectively material world "out there" and our observations lead us "in here" (in my mind) to think about it and formulate theories and understanding? Are "phenomena" *something* independent of me, pre-existing, or are phenomena actually my experience of a relationship between self and world, awakening and coming to light in the perceiving, beholding human being? This is a very tricky, subtle but *very* important question.

STRENGTHS OF MODERN SCIENCE

On the other hand, in a pragmatic view, people set aside the questions of whether the knowledge is "real," and presume the task of natural science is to simply reproduce an image of this layer of reality as precisely as possible, and organize it in a comprehensible way. So, even if a subjective prior conception and historically-related world paradigm play a role, then the image we build up (modern natural science) would have the following strengths:

- The path from theory back to phenomena can be clearly retraced using understandable concepts, as must be since this method deals only with the aspect of phenomena understandable in thinkable, quantitative categories. Thus, this whole construct will be *comprehensible.*
- As we go beyond mere acceptance of an invention and mere application to build up theories of immense breadth at the highest level, daring and bold intuition plays a role. Such ideas open unbounded possibilities of activity and schooling to the human intellect. Thus, the whole is *inspiring.*
- In applying such theories and methods in working with nature, technology arises; we needn't argue its effectiveness. Such a science is *usable.*

Even if the whole range of scientific theories isn't taken as objective general truth, nevertheless most people feel that, in fact, it *does* achieve the best that could be achieved *by* and *for* people, with the means at their disposal. Although it might encompass and weave in more of reality with ever better revisions of its theories, they can't see how anyone could achieve something more understandable, more inspiring, or more effective! Most people concede such a natural science may not be the only one possible, but believe it is the best that can be achieved, and thus should be acknowledged and supported by everyone.

PURE PERCEPTION, A WELL WITHOUT A PIPE

We have explored the conventional ideas in order to contrast our approach of phenomenology or Goetheanism. Our starting point for all knowledge and thinking is actually *perception* through the senses, the active participation and perception through human senses of a living human body. But pure perception is very elusive; we can't actually *say* anything about the senses unless we go beyond pure sensing. We can explore the world by means of perceptions, but we can't discuss the world in perceptions. Nevertheless, we start with perception as a relationship, one that goes far deeper than these concepts we are using to discuss it, and even lies *prior* to them.

But, we can't prove this, since proof already goes beyond perception to use conceptualization. We can have pure perception only when we *hold back* our capacity for forming ideas and concepts. (Steiner 1918) **Only pure perception exists per se—but we cannot say anything about it.** The reality of the way it arises in the human soul is actually something holy; what we could say about the content of perception is it is a human product, always a prior conception and a theory conditioned by the living world.

THINKING REVIVES IN PERCEPTION

A person could object: Then every theory is just arbitrary; but if they are neatly formulated, they relate to the world because they are technically usable. We respond: True. Concepts and theories are certainly added to perception by *human* effort, however they plainly prove to belong to the world if people have made an effort, but correspond to experiences of *a different* side of the world, one not revealed in perceptions. Our thinking is an essential complement to pure perception. In thinking, people have a kind of organ for this other side of the world, which is most active when we reflect on the world of senses rather than the world of thoughts. Goethe, in contrast to many modern thinkers, understood perceiving as a particular conceptual activity.

PARADIGMS AND THE LIVING WORLD

As soon as one transcends this reductionist way of asking questions, and thoughtfully begins to work with the qualitative side of human experience, there is no longer such a thing as phenomena *independent* of theory (conceptually organized perceptions). The phenomena are the aspects of the world we live in through being conscious in perception. In the process of thinking, through a balanced consciousness, out of the experience of perceiving naturally unfolds a definite kind of conceptual activity. The more we live in the perceiving side of this balance, the more the true being of the world will actually speak in the thoughts and concepts which the phenomena lead us to form. Pure perception exists per se; phenomena are already passing over to a more conceptual side.

This is important because the implications are very broad and deep. The reductionist approach, considering mind and self as separate from the world out there, has a split implicit in it: "The spiritual and moral ... dealt with as epiphenomena. ... On the one side is human life with its values and choices, on the other, a scientific Utopia: Man as Machine." (Jung 1981c)

INDIVIDUAL DECISIONS

Since we can prove neither method with logic alone, it must remain a free decision of each person which way they wish to understand their relation to the world: theories about the world which allow one to manipulate the world, or a relationship of perceiving. Basically, we can only say that the author has worked through the phenomena in this way, and *intends* to work with them thus. The Western tradition tends toward manipulation of the nature out there; a complementary approach lives more richly in the phenomena and uses more the subjective, qualitative side of perceiving. It does require a carefully schooled, balanced, thoughtful consideration of the phenomena and of our process of forming concepts.

IV. ALTERNATIVE SCIENCE - NO THANKS?

LONGING AND HORROR

We have reached the most explosive place in our introductory discussion. Even in one of the most inviting works on teaching methods, it can happen that when it comes to the theme of alternative approaches to science, the whole style changes dramatically: Irony, dire predictions and all-out polemics surface. This *although* or perhaps *just because* alternative science increasingly becomes the vague object of longing of many people.

The editor of *Chemistry and Technology News* proposes, "Do you believe that it would be possible ... to develop a rational scientific method which doesn't pursue just this reductionist approach to nature, rather a method which sees as much of the whole and simultaneously isn't useful technically? You say yourself that's a royal task. ... Couldn't we at least strengthen a consciousness that there exists a natural-scientific-technical world conception other than our own? Are there certain sciences which are not intended for utility, but earlier on proceed in a phenomenological way, and if there are, shouldn't we pursue them?" There is nothing to say but YES!

Another author expresses alarm that "science is not features of rationality or freedom, not basics of education; it is a commodity. The scientist himself becomes a salesman for this merchandise, they are not judges about truth and falsehood. ... The starting point is not truth, or the newest level of science, or some other empty generality; the starting point is the equivalence of all traditions." (Feyerabend 1981)

The anti-science effect, which has grown so strong today, arises not only because of ecological, political-social crises, but is due to a split in the living stratum of the soul into calculable scientific concepts on one layer and feelings and moral (or animal) impulses on another, which directs practical dealings in life, and yet becomes ever more isolated. It is also on the rise since people no longer hesitate to speak about this split out of "respect" for the acknowledged science.

HUMAN POWERS

In the last centuries the scientific view of nature has established only what the Cartesian revolution, the methods of Galileo, Newton, etc., contain: a kind of imperialistic knowledge which serves the orientation of the human being toward control of the cosmos. The reason the causal-analytic approach to natural science requires an alternative is not because it provides too little *information* about the world, but rather because it doesn't activate our soul-spiritual forces.

WEAKNESSES OF PHENOMENOLOGY

According to the above ideas, there are three things phenomenology cannot have: It is not understandable in layman's terms (in concepts which are definable in a mechanical framework); it has not progressed very far nor inwardly developed (doesn't have 400 years behind it); it is not usable in a naïvely unconscious way, in a mechanical sense, without a look-back at the whole situation in nature.

The most negative is a lack of external provability, without a mechanistic mode of thinking. This is unavoidable, since this provability arises from just those conceptual tools which now should be extended and complemented, namely the quantitative approach, and ultimately mathematics.

EXTENSION OF BOTH METHODS: PROPAEDEUTICS (LIBERAL-ARTS INSTRUCTION)

Alternative and rule-based science complement each other, and neither possesses all the necessary aspects for education; we can dispense with neither one. Even where we can explore with a quantitative approach, we must go inward.

For example, in the gas laws, if we wish to really penetrate the matter, we must go inside the mere formula $V_t = V_0 (1 + 1/273\ t)$. As the heating is increased, the increased tension in the gas container (pressure) is experienced as a reflection of the activity of heat and the increase in measurable volume.

Analytical thinking initially creates distance; but then, an inner connection with the processes in the world, insofar as the magnitudes in the formula we work out are inwardly felt and experienced. Such a going inward is the starting point of all phenomenological research.

If the advanced technical material presented at the introductory level of a broad, liberal-arts curriculum should be limited, what should be correspondingly expanded? This leads us to a human basis for a new teaching method: basing the themes on the development of the student. The questions then become: In which grade should physics start? What ideas should it pursue?

V. DEVELOPMENTAL PSYCHOLOGY:
TRADITIONAL DIVISION OF PHYSICS-MATERIAL

In the pedagogical works of Rudolf Steiner there are many places where he points to the essential reorientation of the lessons at the onset of pre-puberty. He indicates how, mainly after 11²/₃ year (6th grade), the young person separates out the living and ensouled from the dead aspect in the surroundings and is now able to recognize it. Therefore, we can be attentive to this moment, understand why the study of mineralogy and physics can begin then, and relate to the young person's developing natural interest in life.

CURRICULUM OVERVIEW

In 6th grade—when the physics studies begin—the way of regarding things is still phenomenological and imagistic in all the topics. It is not yet abstracted in the sense, for example, of deriving general Laws of Nature. The material-causal method of school physics is kept out. However, in the **7th grade** this comparative-imagistic approach already receives a new direction. Certainly not toward scientific models, atoms and the like, but rather toward work, livelihood, trade connections, and thereby to technical applications in life. Since young people will at puberty begin to distance themselves from their parental home, they are occupied with the question: How can a person help himself along in the world through clear ideas? And, underlying this: How can a person contribute something of value in the outer world of work?

At the end of the 7th grade physics studies, *as* an essentially technical topic, we present the mechanical theories involved with the use of levers. Simple experiments, scientific systematics and a technical-practical understanding of the apparatus all occur together in mechanics. Out of such a treatment, we reach the starting point of classical physics. However, a systematic treatment of the other topics in physics founded on this does not quite follow yet. Even in **8th grade**, we will use quantitative or expressible formulas only in particular cases—for example, in the treatment of current in a circuit or in the pressure calculations for fluid mechanics of air or water—but they will still be connected with direct observations in the classroom.

In 9th grade after the material-causal explanations of the telephone and locomotive have been treated thoroughly (for example, with current-time diagrams, vapor-pressure curves and thermal mass comparisons), this phenomena-based method arrives at a quantitative systematics for the first time only in the **10th grade.** There, even while quantitative, the view shifts back to the human being: How we predict through calculations of a parabolic trajectory is now thoughtfully considered as a phenomenon of knowledge. And, only after the question of perception and reality in the "supra-sensible" is worked through in the **11th grade** via the study of modern electrical inventions (Tesla coil, Roentgen rays, radioactivity), do the light and color studies of the block in **12th grade** penetrate again to a Goetheanism. However, the methodology [and epistemology] is now clear to the students.

So, moving toward the 9th grade, the curriculum increasingly lays aside the phenomenological approach, as we increasingly lay aside nature. With the analytical method and use of technical instruments, a non-spiritual, material causality chain now becomes the goal of the lessons. This is what is appropriate here and therefore—pedagogically viewed—nonetheless a Goethean method (Goetheanism in the consideration of a thought sequence). From these technical-practical topics of the 8th and 9th grades and their social implications, we turn around again in the 10th grade and focus on the thinking human being. The process by which we develop knowledge and the relationship it has thereby to the world is raised to the phenomenon here. Through such a careful and interconnected arrangement of steps in physics, the young students gradually awake to an exploration of their own methods of knowing—and that is the real question.

SUMMARY

The so-called abstract study of natural science, using causal-analytical methods, actually lies near other paths. (see chart next page) First we can consider the overall experience reading downward in the central column: A restriction to increasing abstraction and models, providing the all-permeating meaning of the world, and degrading the initial holistic experience to a misleading reflex, a byproduct of the model (an illusion). Knowledge becomes imprisoned in the lower circle, and people run around in a senseless experience of a causal-analytic mechanistic world. They develop gigantic technical works and enjoy everything that they love.

In contrast, working upward on the right is a deepening of a phenomenological way of considering the world, leading people to truly grasp who they are, and thereby building a basis for a new approach to the spirit—a holistic experience. They enjoy what they *realize*. With this we note how the 6th grade begins with holistic knowledge; the 9th grade moves to causal-analytic thinking but not yet models; and mechanical thought in 10th, where we stop.

TWO PATHS OF NATURAL SCIENCE CONTRASTED

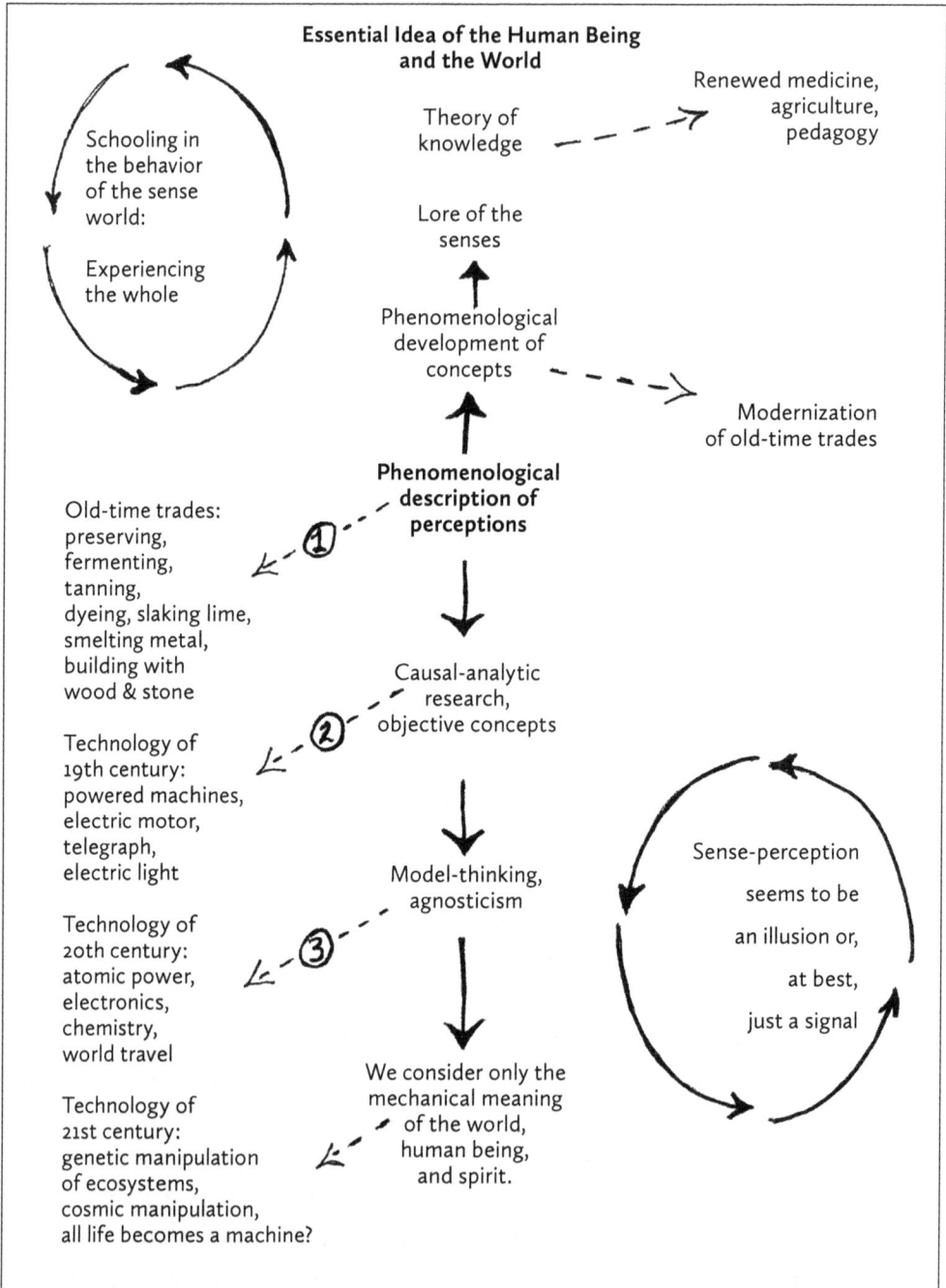

Essential Idea of the Human Being and the World

Theory of knowledge

Renewed medicine, agriculture, pedagogy

Schooling in the behavior of the sense world:

Experiencing the whole

Lore of the senses

Phenomenological development of concepts

Modernization of old-time trades

Old-time trades: preserving, fermenting, tanning, dyeing, slaking lime, smelting metal, building with wood & stone

Phenomenological description of perceptions

①

Causal-analytic research, objective concepts

②

Technology of 19th century: powered machines, electric motor, telegraph, electric light

Technology of 20th century: atomic power, electronics, chemistry, world travel

Model-thinking, agnosticism

③

Sense-perception seems to be an illusion or, at best, just a signal

Technology of 21st century: genetic manipulation of ecosystems, cosmic manipulation, all life becomes a machine?

We consider only the mechanical meaning of the world, human being, and spirit.

VI. PEDAGOGICAL METHODS

Nowadays, natural science is pursued with the goal of bringing facts to light and utilizing them technically. Whether research makes an impression on the public depends on the usefulness of these facts and on what can be produced from them. The analytical method is impressive and shows the basic significance of the things investigated. But, how do things stand with knowing?

In order to unify and order the innumerable facts, physicists build theories and conceptual models, e.g., the wave model of light. They know, in principle, that every model has limitations and that above all there are always phenomena which will contradict it. A model applies only to certain phenomena; for others, one must construct a contrasting model for the same topic. So it is that, admittedly, no knowledge of the whole arises. Nevertheless, scientific models such as the light theories or the atom-concept (which never touch the essence of the matter) are commonly viewed as the essence, or at least as steps to the essentials—as much in the popular view as in scientific consciousness. In thinking this we overlook the fact that questions about "essence" fall entirely outside the categories of physics, cannot be included in them, and equally cannot be answered by them. Nevertheless, to the student every model is naturally taken as an answer to essential questions, a statement about the ground of existence, although the students are not consciously asking about it! The narrowness of these kinds of answers cannot really be appreciated by the student, since s/he does not yet comprehend the role played by pure thinking in forming world-conceptions, or the way a free activity of the human being is connected with it. The student is oriented more to perception.

Abstract conceptions, however, which do not originate from the phenomena, become misconceptions: as if not invented by us, but rather as coarsely material or also as magical beings which we light upon and then believe that they stand as final cause behind everything there is. And so the phenomena give rise to an ontological misunderstanding of physics [misunderstanding the nature of being or reality].

Goethe's natural-scientific work, as set forth by Rudolf Steiner, developed a little-known qualitative way of observing nature and now is constantly producing and extending a literature of Goetheanism and phenomenology. This concrete path built around an intercourse with qualities of perception will actually bring the perceiving human being to reality. Due to the style of teaching typical of the later elementary years, there is a fundamental need for qualitative natural-scientific methods, since only these make an attempt to work in accord with the essentials, to work on just those questions deeply experienced at this stage of life. The question of whether such an approach promotes complete understanding can remain open, but certainly it is more complete than that promoted by models. For us, the pedagogical value of the teacher's striving and his inner, spiritual activity exceeds the value of mere knowing.

TWO PATHS TO NATURE

This suggests two methods of conceiving nature (and thereby also the world) which at first are unconnected, namely: (1) the material-causal method, which wants to find a model of the ultimate cause through analyzing matter, which is thought to lie at the foundation, and (2) the Goethean method of phenomenology, which seeks to order rich perceptions of nature as a totality. These two ways of approaching nature are both contained in Dr. Steiner's curriculum. However, he didn't intermix them, nor did he present them interspersed with one another. Rather, we find them organized into the curriculum with the analytical following the phenomenological—the aim being to place them in an evaluable contrast by the 12th grade.

VII. TO THE TEACHER

HEALTHY AND UNHEALTHY PREPARATION

This book has been written precisely for those who won't receive it first: the students. Teachers will doubtless receive it, perhaps with a sigh of relief. It may seem that in it lies a wealth of physics material, neatly packaged and awaiting use like pre-assembled concepts, stashed away in a deep freeze. The impression that I have carefully tried to create, to be understood perhaps only later, is that such a hope cannot be realized through a book; and if it could it should be forbidden! The students do not live off the stuff from a book, but from the initiative of the teachers and from their spiritual wrestling and persistence to blaze a path for them on how to see the world. As the teacher learns, so the students learn. Whoever is in good command of the subject—knowing that she must not only freshly order things but must unlock new sides of it *for herself*)—will be able to teach out of direct, spiritual experience—she will inwardly reach the class. In order for the physics covered here to encompass this dimension, it was necessary on the one hand to make the subject easy for the teacher, so it could be expressed in words, and on the other hand, also necessary to make it hard for the teacher. This challenge comes from me, since I can understand the full scope of the material only in certain scientific and pedagogical terms.

Following Goethe's beginnings and in the spirit of Dr. Steiner's theory of knowledge, a person should *attend to nature's phenomena and simultaneously attend to the inner activity and experiences of the [perceiving] human being;* then the person must attempt to break free of the rigid [mental] labels of "particles," "waves" and "energies." New, basic ideas are of great value to go more deeply into nature, but these are radically different from the traditional concepts of school physics, and from a materially disposed, conventional understanding.

My ideal would be to set aside the quick definitions in phoronomic terms [purely mechanical motion], and lead everything over into open, seemingly indeterminate concepts. What light "is," what sound is "in actuality," what is "the basis" of heat—all this cannot be clarified without further ado, but must be achieved by experienced observations and thinking on the part of the teacher and the students. We call upon the reader's own perceptive judgment. (Compare Goethe's concept: *Anschauende Urteilskraft,* "perceptive power of beholding.")

Clearly physics classes shouldn't be merely a list of points out of the great book of scientific authority, but should be spontaneous, without limits at the outset to questions involving only quantifiable measurements. Naturally, this requires a lot, namely: **an alternative way of thinking,** the greatest possible individual sense of orientation, and **a familiarity with the phenomena.** Only to the extent that this text seeks to pursue such difficult goals, should it be allowed to appear. An easy, practical collection of obvious school-knowledge—provided it is good—is only a help to the teacher who is already in desperate straits; however, it undervalues just these deeper needs of the students.

Aside from aiding such a direct, untrammeled, personal spiritual quest for knowledge, teaching has, naturally, a more ordinary side: We also must furnish the students with actual knowledge of the thinking and the day-to-day business of our technological civilization. For this the teacher needs information which is refined and prepared. So I have interwoven a few such indications in the main text and also provided some related topics. There we can see how we can work without abstract scientific models and hypotheses so far removed from the phenomena.

EXPERIENCES WITH THIS METHOD

Interesting teaching experiences have arisen meanwhile from colleagues who have received this book just because physics was new for them. In the worst cases, the students said "That was, certainly, all very logical," meaning that the phenomena were already clear (they weren't led to previously unknown phenomena). On the one hand, they were caught up too little by the wonder of the world; on the other, not enough difficult questions were given them.

Now, this raises several issues. The students have a point: New facets of the world must be shown in wonder-filled, previously unknown effects. The students like to plunge into manifestations of the lively activity of strange forces, they like to see things which can be conjured forth with simple manipulation. Whenever the teacher surprises them with unexpected phenomena, that makes an impression! The manipulative, perceptible side of the world must come to the fore. True, without a wakeful musing and reflecting over the experiments, a certain superficiality and inconsiderateness would be fostered. Nevertheless, I do a few direct, surprising presentations. But, in caring for the perceptive side of things, one should not neglect the thought-aspect.

To an art of experimentation belongs also an art of conceptualization. A person will place himself perceptively into everyday phenomena and, out of his own careful pondering, form new concepts, so as to build wholly on the phenomena. Then experiments which appear commonplace will no longer seem merely logical but, perceiving with new eyes, will allow the thinking to reach up to the working ideas of nature so they really speak. (see Goethe's essay on plants, regarding his "seeing" the archetypal plant) It shouldn't happen—as it often does in conventional physics—that the teacher simply presents how things are supposed to be seen. Goetheanism as pronouncement remains mere semantics.

Conventional physics, on the one hand, relies on everyday thought-forms which have a life of their own and therefore easily take root in the students. On the other hand, the more the teacher has the foundations of Goetheanism in herself, still utilizing and experiencing it even on days when she doesn't teach it, the more she is able to call it forth out of the simplest phenomena and give it a permanent basis in the students' souls.

Using a book to show the way to do this is, naturally, risky. In the worst situations, the impression could appear that the proposed material is too childish and that the 6th grade material should already have been done in the 5th grade as it lacks causal thinking. By all means, twelve-year-old students (grade 7) should already practice a thinking which takes its stand in the compelling links or relationships between specific phenomena. We only have to try and develop a thinking *freer* than a thinking which explains the appearances out of processes of the matter which is supposed to make up the substratum of reality. This will be shown particularly in the section on light.

UNUSUAL LANGUAGE

It is necessary in the 6th and 7th grades, and somewhat in the 8th, that the students immerse themselves in the qualitative aspect of things. Thereby, their experience receives an orientation toward objectivity, not through analytical abstractions, but in careful expression and in connecting to the phenomena. If one wants to work in this way, one needs *expressions* and *a language* which embraces *the feeling side of experience.* The one-sided cultivation of the cognitive faculties, basically only rational instructional elements and goals only in the cognitive domain, fails to take into account and neglects the feeling and doing side of the human being. The student learns individual pieces of knowledge. But the feelings will not be expressed, developed and differentiated: "The feeling of senseless learning becomes often a feeling of senseless life, the seed of a depressive orientation." (Fischer-Wasels 1977) Through abstract teaching, the desire for an emotional reality can lead to a splitting of personality. As in many such articles, a rigorously restricted subject matter is called for and, for the natural sciences, a strong consideration of the phenomena. [see in the U.S., the move by the PSSC curriculum project to emphasize simple phenomena with simple apparati. – Transl.]

As experience shows, such a *restriction* of the science content will be of little help if a *new art of thought—a different language* and an experiential way of working connected with the phenomena—does not simultaneously replace a mere taking up of one phenomenon after another. One can only put *something else,* not nothing, in the place of an abstract thinking that has become a scientizing. This "something different" should engage the complete *experience* of the human being, not only our *conceptualizing* faculties. That is to say: In the course content, new fundamental concepts are necessary! The technical language doesn't suffice.

What is attempted in this book is to utilize such language as *unites* the feeling with the conceptual. But, this is the main snag for professionally trained people. Rather than an illumination of actual perceptions, they expect to combine definitions and concepts purified of every experiential aspect (e.g., *phoronomy* or kinematics; motion without regard for the actual units). Already on pedagogical grounds, we cannot here concur with this goal. Thus, we must attempt to intersperse these rational and emotional instructional elements, not only like a spice, but rather to draw them right out of *the things at hand.* From the most fundamental perceptions to the most all-encompassing thoughts, it is important to always speak to the whole human being and to truly look on concrete phenomena of the world right through to the end. That leads necessarily to unconventional, open-ended concepts and indications.

We take into consideration that these new, open-ended fundamental concepts have their source in perceptions saturated with feeling, they lead on to meaningful thoughts, and they have a structure which can be evaluated as a result of their relationship to actual perceptions. We aren't dealing with a complex of feelings arising from some kind of subjective evaluation, but rather with elements of science, [a conscious process]. The full earnestness of the task of physics teaching, as presented herein, will work as a stimulus when one has the insight that the central concepts we strive for aim not at a childish prettying-up of the barren subject matter of nature with all sorts of subjectivity, but rather, our aim is to open up a new scientific reality out of our own personal perceptions—however modest the beginning introduced here might be.

You might notice that the new concepts in the section on light are a bit further developed toward a Goethean phenomenology. With heat, especially in grade 6, such a Goethean approach would be impressive but also becomes more difficult to grasp. Acoustics in grade 6 is also organized around unconventional ideas, and only in the subsequent grades will become more familiar. But in acoustics, these new concepts do not take such an unusual form as in light studies. Electricity and magnetism are presented in a qualitative way and lead to concepts formed in a particular way, though these remain mainly empty forms, as yet unfilled with much that is compelling or germinal. In mechanics, least of all have we made a beginning into an experientially clear treatment of force and pressure (hydromechanics).

We are dealing here with a search for an understanding that takes as its starting point the human being. In order to live into such a new mode of approach, the epistemological orientation must be spoken of in an extensive work, amply supported by philosophical and anthropological points as are here only begun. And the physicist would nevertheless still not be free of a deep uneasiness—for, instead of studying the rewarding scientific developments of this century, we have apparently given them up in favor of a new consideration of Aristotle. Whether this is necessary, something to put up with, or just foolish cannot be determined from concepts but only in one's own intimate meeting with the perceptions.

If I occasionally draw on concepts from the writings of Dr. Steiner, this should provide those familiar with his work a quick means for further reading. However, the seeds of the phenomenological method and what is therefore useful in teaching are not dependent on these references. Thus, I hope also to invite many useful contributions from the non-Waldorf teacher.

THE BOOK'S LAYOUT AND ABBREVIATING THE MATERIAL

Concerning the **structure of this series of three books for grades 6, 7, and 8,** may I say that it follows the sequence of grades 6 through 8. The experimental descriptions are, with some exceptions, extracted from the text and inserted after individual chapters. Appended in Book 3 is a list of equipment and, as a possible deepening, a few supplemental topics.

The extent of material presented often exceeds what can be handled in one block, especially if a teacher is presenting this for the first time and isn't at all familiar with many of the phenomena. However, one should not omit an entire topic, as for example, the one usually coming last: magnetism. At least a central experience from each topic is desired.

In attempting to **shorten** this syllabus, the teacher will have a hard time, as I have attempted to build up each topic carefully and thoroughly out of the simplest phenomena and thus establish a line of thought to foundational ideas. Thereby, a certain systematizing has entered in. Each proposition is prepared for and also proven from the preceding one. Such a structured form is indispensable for the teacher in order to introduce him to a new way of seeing; otherwise, he would have no solid ground under his feet for further questions by the students. He must be able to base his conceptions, carefully considered, on an extended field of perceptions. It's different for students: They certainly want to evaluate and establish, not out of a carefully controlled process of construction, but out of insights flashing up out of the circle of their own experiences. They certainly want hints and, upon occasion, an awareness that the concepts taught them are good judgments and are well-founded. But, they don't seek systematics or proof, rather a thinking which is experiential, primal and spontaneous—not a cornerstone for proof. Of course, proof must be given and the teacher must know it, but he should not lecture incessantly afterward.

Thereby it is clear how we can abbreviate: Dispose of systematics for the teacher; dispose of half the experiments. What appears to the teacher as most graspable, most characteristic, remains as a residue. That which is omitted can be recalled in great sweeping brush strokes and summarizing thoughts. For the few students who desire a definite systematic structure, or who want material for knotty reflection and love abstract summarizing-thinking, a great deal more of the systematically based material can be woven in on the side. With experiments, one should not bring in too much nor omit too much. For each experiment, there should be time to lovingly work it up, to present it in its fullness.

Grade 8 Overview

We should not cover the topics in the same order as in 7th grade. It is best if we begin with **optics**. This topic is the most phenomenologically worked-through. With the next topic of **heat**, however, a certain materialistic quality is visible here and there, for example with heat-convection in air and water. This is in contrast to radiation.

The next topic is **electromagnetism**. With the magnetic field surrounding a current-carrying wire, the 'poles' we learned about in grade 7 have disappeared. The 'ends' of the bar which bounded the intervening space where the field appeared in 7th grade have moved away; space itself is active in an unbounded way as the circling field, as directionality. With the magnetic coil, e.g., in relays, as magnetic forces bring the hinged contact into place against the anchor plate, technical apparatus begins to function. Again, we have a mysterious in-between space. And, the telegraph overcomes the space of entire continents, where the earth itself is used as one conductor in-between, again in a mysterious way. What happens in these in-between realms? That is the theme of this block

More graspable is the in-between space of force transmission through water by means of pumps and flasks in the **hydraulics** topic. Force sprouts in a material way from place to place and collects in the depths of the water column. In **aeromechanics**, the space is open at the top, and low pressures and vacuum are experienced. The immaterial 'force of sucking up' is flipped over to its opposite, in the force coming from the other side (the atmosphere *pushing up*); the space between is closed up (e.g., in the barometer). The most strongly diversified in-between space is in **acoustics**. Rarefication and densification move outward in waves from an impact (a 'bang'), creating spherical waves, reflecting, etc. Thus, the block goes from intangible optics (where the in-between space is not interpreted using rays) to the manifold wave shapes created in the in-between spaces in acoustics.

Thus, we have the sequence: Optics > Heat > Electromagnetism > Hydraulics > Aeromechanics > Acoustics.

In the 8th and even more in the 9th grade, the entire way of dealing with the topics becomes more and more similar to a conventional study of physics. Objective relationships are investigated. To some degree, we even seek out material causal chain of events in time and space; the individual phenomena are led over into processes in self-existent matter which do not appear along with the phenomena— for example with waves in air in acoustics, or with convection in heat. Nevertheless, in teaching we should take care not to speak in such a way as though sound was nothing more than alternations of air pressure, or the atmosphere nothing more than a dead substance filling space, acting under mechanical pressure, or as if heat were nothing more than a transport form of energy.

We need objective analysis of phenomenal relationships in order to be able to present sufficiently challenging exercises for the students, to facilitate their developing independent clarity about what happens or doesn't happen, so they can exercise themselves against external fundamentals. Teaching what is merely descriptive natural history, even if it tries to explain the deep essential connections, is no longer productive. We must give the students something to bump up against, and the external might of the world must be experienced.

Already in the 8th grade, it can be desirable to pursue a topic into its technical applications. The students should be able to experience how things relate in reality, how they stand in real life. For example, we could visit an electric street-car repair depot to explain motors, resistance, induction and the like. In a similar way, we can introduce the especially rich supplementary topics of the telegraph, air pressure, and altitude sickness. These bring experiences from outside, from the widths of continents, and make a valuable complement to the apparatus of the physics lab.

Grade 8 Optics

Since we will not deal with optics again from 9th through 11th grades (only again in grade 12), a great step must be taken in grade 8, a kind of 'telescopic' step which opens up the distances to us.

I. METHODOLOGY

We deal with two aspects of optics: (1) geometric-spatial and (2) color.

Optical magnification by prismatic refraction can be studied impressively in water, as for example in the magnifying-glass effect of a dewdrop, or as seen through water as it flows over a stone or in standing waves formed in a ditch. But, this occurs only with adults, or with 6th graders. It presumes that a person will 'offer up' his thinking, that with their own forces they are able to move actively in thought among the phenomena, allowing their concepts to arise as if sculpted amidst them (adults); or, yearning to be active, totally go along with the lead of another person (lower elementary students). In the former case, the 8th grader can't do it yet; the latter, he doesn't want to follow any more. *They don't think selflessly, but rather more subjectively—making judgments which are sometimes rash and sometimes reasonable.* In this way they seek out the raw material of experience in order to try and grasp larger problems. We only have to take care that they develop their own thinking. This will be served when the world around them makes subjective impressions, yet also becomes problematic [forcing them to think (objectively)]; thus, questions dealing with a subject have to be formulated in an explicit and clear way, but also as deep questions.

These questions ought to deal with the many consequences of how people manipulate phenomena, not so much with the delicate natural order -this way they will interest the 8th grader. Thus we can describe the 8th grade style of learning. We start with experiments which are clearly laid out, and concretely discuss optical apparatus capable of creating new views of the world. So, the finer natural phenomena are placed at the conclusion, so that the weaving tapestry of nature—penetrated with thought and will—is finally experienced through feelings of beauty. An approach, centered on apparatus, will usually appeal mostly to young boys. Thus, for the girls, we can thoroughly discuss and include anecdotes of, for example, how people experience near-sightedness, what problems they will meet and how we can help them, i.e., *present the purely descriptive in its experiential significance in life*. Perhaps we could test a near-sighted student or else describe such tests in connection with the biology blocks. The usual approach, explaining short- or farsightedness by the accepted theory of ray-optics, should be avoided; such a view summarizes near-sightedness only as a problem of ray-paths in a vitreous body badly shaped by mere chance, and not as an external expression of the personality. Also, we should not impose a theory of optical-defects, but rather *describe with empathy the universally human experience of vision*. This would balance out the sharply conceptual and experimental-literal descriptions of sight given in other parts of grade 8 physics.

This objectification by experiments can also be compensated by something else, although we should be clear from the outset that objectifying things is therapeutic in this life-phase—so long as it doesn't go so far as to make them rootless; the students should find their way to the outer connections existing in the world. In natural science teaching this means: hearing what a person has never heard; seeing what they have never seen before; experiencing in heat what they have never experienced there before, etc. If we can really observe this orientation in the students, then we can realize what it would mean for them if we began the optics in a meeting with nature or else built it up from the most grandly developed experiments. Yet, in order to *experience something new in the meeting with nature requires selflessness and consciously-directed spiritual forces*. Also, if the teacher fully understands ray-optics, then they know it would only partly correspond to this stage of life. In contrast, exploring the facts themselves as they unfold in a new, impressive way, helps redeem the class teacher's one-sidedness. **"Behold! Evaluate It Yourself"** stands over the archway to the upper classes.

In the 8th and 9th grades, more than before, we need impressively developed, even focused experiments; nevertheless, at least in 8th grade, this must be restrained a bit so that an effect can be built up, can still be called forth by the next one, so subsequent phenomena will be reinforced by a surprise effect. We could go from the experiment to the related gesture of natural phenomena: in the beginning of the experiment we could illustrate what we see under water (see below). Thereby, we would already bring into the experimental phenomena *a delicate feeling for the physiognomy of all of nature*. Later on, this can be deepened further when we go into clarifying what we ourselves experience through our senses. For example, when we look into deep water, does a person simply get drawn down into the underwater **space** with his experience, or does what we see remain a planar image?

What would be required of me down there? Or: who has trouble with the pincushion distortion of the images of our lens (as does happen! vertigo?) What is unpleasant about this? Once again, we are dealing with the experience of the **whole** human being.

II. THE VIEW INTO WATER

"LIFTING" (GEOMETRICAL-SPATIAL)

Everyone knows: viewed from near the ground, the bottom of a pond always is deeper than we would naively assume it is. Specifically, a bowl filled with water (distant enough so we can see the ground under it) appears to be flattened—the bottom appears right up near the water surface. It is not possible to view a deep, broad expanse under water while maintaining a dry vantage-point, the way we could survey a valley floor from up in the air. For water, we can see into the deeps only relatively near our standpoint. If we look diagonally to a more distant place, aside from the mirroring studied last year in grade 7, the water appears flat and the bottom somehow seems close by. As in the Jack London story "The Alaska Kid, the Man on the Other Shore," we imagine two gold-prospectors on the legendary "fabulous sea" under the Alaskan glaciers. Each sees Faustian gold nuggets at the feet of the *other*, seemingly only in knee-deep water, whilst the water in front of their *own* feet appears meters deep. Each waits in vain for the *other* to pull out something. Or, how remarkable is the masterful performance of the kingfisher bird (daily coping with such refraction problems).

Such phenomena are seen in a fountain, pool, bathtub with a flat bottom, or even in a basin full to the brim. In a forest pool or at the river's edge where a shallow slope allows us to stand down close to the water, a submerged branch a few meters long seems to swim just a hand's breadth under the surface, hovering between objects floating on top and things resting on the bottom. If we look from as low a vantage point as possible (quite obliquely to the water) the branch is no longer visible. Where we stand is always the spot where we look into the nearby water most vertically and there it appears the deepest; as we walk along, the bottom surges deeper. Another field observation can be made on a branch which sticks diagonally out of the water: it does not appear to go straight out when we look at it from above. Especially from the side, it appears bent; it is always the immersed portion which appears bent upward.

For the reasons mentioned, we should show such phenomena initially thoughtful arranged, and in a pure, *experimental* way in the classroom. An aquarium serves this purpose quite well, or our lab tank even better (Op1). We find: The more obliquely we gaze at the water's surface, everything we see is somehow lifted up.

We can ask the students how high up the far, lower corner of the aquarium tank appears as seen though water, from where the corner seems to be by touch on the outside (Op1b).

Visible depth is much shallower than the tangible (what we can touch). If we look perpendicularly into the water, we notice that a slight lifting still persists (Op1). An example: The tank is 28 cm deep (~11"); at the bottom lies a penny. It appears a certain size as another coin alongside the tank, out in air. However, the one *outside* would have to rest 7 cm (2.7") *higher up* than the one *inside* on the tank bottom to *appear* at the same depth. The coin inside appears to lie only about 21 cm deep. This is the *visible depth*, D_V based on the visible size of the object. The tangible depth, D_t, corresponds to the 28 cm deep tank. The relationship between the visible depth, D_V (the theoretical distance to the object which would produce the observed size) and the tangible depth,

D_t leads to the so-called refractive index: $\dfrac{D_t}{D_V} = n$

In our example:

$$D_t = 28 \qquad D_V = 21 \qquad \text{so } n = 1.33$$

The index of refraction (refractive index) n is a dimensionless number, always greater than 1; the refractive index for water is 1.33. A person looking vertically in a water basin sees the bottom appear "lifted" about ¼ of the depth, and thereby a bit enlarged; the water appears about ¾ as deep as before.

The refractive index is different for each fluid; water has one of the smallest. The larger the value, the greater the so-called "optical lifting," especially for an oblique viewpoint. Conventionally, this index is measured by observing how a beam of light (considered as a light 'ray' or 'ray bundle') is "broken" at the surface; the ratio of the sine of the angle of incidence and sine of the angle of refraction (both measured from the "normal" or perpendicular to the respective surface) gives the *refractive index* as first worked out by Snellius in 1620. Our explanation, based simply on the ratio of tangible/visible depth, can serve in 8th grade since the trigonometric sine has not yet been introduced. This approach is presumably more effective anyway, as it is based more directly on the phenomena.

TABLE OF REFRACTIVE INDICES	
Material	Relative Index
Methyl alcohol (wood alcohol)	1.27
Ice	1.31
Water	1.329
Ethyl alcohol (grain alcohol)	1.36
Acetic acid (vinegar)	1.37
Glycerin	1.45
Linseed oil	1.48
Benzene	1.50
Turpentine (pine oil)	1.51
Cinnamon oil	1.59
Carbon disulfide (solvent)	1.63
Methylene iodide (diiodo methane)	1.74
Fluorite	1.43
Quartz glass (Pyrex)	1.46
Window glass (soda lime glass)	1.47
amber, natural quartz	1.54
Beryl	1.58
Ruby (corundum)	1.77
Zircon	1.96
Diamond	2.42

Some semi-precious stones are included here for comparison. Unfortunately, such numbers barely enable us to visualize the experiential depth relationships (as defined in the formula, above).

The beauty and glittering quality of gems and their play of colors demonstrate two additional effects, also exhibited by water to a slight extent, and connected with this 'lifting' (refraction). First of all, is so-called total-reflection: When we look very obliquely at a water-air surface (from under the water), the surface reflects like silver. Also, we are familiar with the silvery sheen of bubbles seen in a beaker of boiling water. (Refractive index will be studied more precisely in grade 12.)

COLOR

The second main phenomenon is color fringes (Op1). The more oblique the viewing angle, the stronger they are. With its high refractive index, which acts like a very oblique view, diamonds can sparkle with myriad colors. So, how can we explain their beauty and colors? Actually, they arise in accordance with a Goethean archetypal phenomenon. Where a bright scene is made darker, e.g., by milkiness or turbidity, there we get the warm colors: the red setting sun or yellow snow-fields showing through the haze of distance, for example.

If a dark region is brightened by an illuminated, turbid media in front of it, then the cold colors arise, e.g., azure sky; bluish distant mountains; thin, light blue wood smoke in front of a dark, tree-clad mountain. We can investigate this phenomenon with an experiment (Op2) which shows how this effect not only depends on the alignment-of-view (e.g., hazy air over snow), but also upon the obliqueness of viewing angle.

A small spot of darkness made by the forest in my visible hemisphere (cupola) is also surrounded by bright-white sky. The turbid smoke draws in illumination from this surrounding brightness and acts as *light in front of darkness—thus blue arises*. Also, in this situation the slightly turbid air is *illuminated from the side*.[1] Thus, the contrast becomes moderated—the way dark tree branches seen in a misty forest seem less dark. Turbidity or milkiness means that *something of the adjoining scene is interwoven* with the scene in the direction of view, producing a brightening or darkening. If this interweaving is sufficiently delicate, then colors arise: the violet-blues for brightening, the yellow-reds with darkening.

If a dark strip is optically shifted away from the observer (stretching and elongation) by the refraction of water in a tank, then one spot (one particular direction of view) which was previously bright is now dark [*darkness shifted over light*]. So, with this *refractive darkening, a red-yellow color occurs*—just as with transmitted or interwoven darkening (see smoke before the sky, above). Since this change (darkening) does not occur everywhere on the viewed surface but only at a boundary or edge ('lifted' by water contents), the color change is restricted to this boundary [*colors arising at the meeting of light and darkness*]. But the colors persist even when we aren't filling the tank further, when we aren't moving our head lower (which 'lifted' the view of the tank bottom)! Are the colored boundaries just a residue of the displacement which created them and now seem to have ceased? No, *the colors exist in something objective, namely the split between the visible and the tangible position*—whether stretched or squeezed, each visible object is always displaced from its tangible position. The concept of milkiness here metamorphoses from something objective (a haze due to droplets) into something functional (something inner, experiential—a way of thinking). The colored regions are a persistent expression of a temporal drama, which they still show in the resting image. There, they manifest the turbidity of something nearly transparent, but changed, transformed, now altering [sundering] the visible-spatial association of things.

III. THE PRISM

An apparatus which only lets us look diagonally inward as well as outward is the prism. It presents an erect water surface, behind which is air once again. We have a student stand a few meters behind a large water prism, but slightly to the side from the direction of view so that he does not stand directly behind it (Op3). Everything seen through the prism will be shifted to the side:

Figure 1. Displacement in a water prism

Seen through a water surface, an object is *'lifted' in the direction my gaze would shift if I looked more obliquely* (less vertically) toward the water surface; it is the same with the prism. The scene is displaced toward the narrow edge, the acute corner (the illustrated prism has its acute corner to the right, the double one to the left). And, if we make the angle at the acute corner larger, the displacement gets greater (Op3). It is as if our gaze slid off the slanting surface (whether water or glass). We have already established this with water; if we try to estimate the depth while looking diagonally, then our view 'slides' away on the water surface, slanted to our view. Moreover, the more slanted the water's surface, the less we can see into it, until finally we see only something reflected and can no longer see into the spaces of the water's depths.

Looking up out of the water, or looking perpendicularly to the front surface of the prism, we must say: The view seems to be drawn in, to 'slide in' to the inner surface of the prism. However, this is a case we will defer until a more detailed treatment of refractive indices and Snell's law in grade 12 optics.

We now allow as many students as possible to look through a small water prism (Op4). Usually, they will see a strongly displaced colored band (on dark edges with strong contrast). The rule of displacement toward the narrower edge (conventionally: the most acute corner) can also be observed in its variations by holding a glass prism point-upward or point-downward or in other directions and observing the colors which arise. With larger prisms, such as in binoculars, when looking very diagonally, refraction and distortion also occur; however, we will not investigate these aberrations further at this time, but will focus on a transition to the lens.

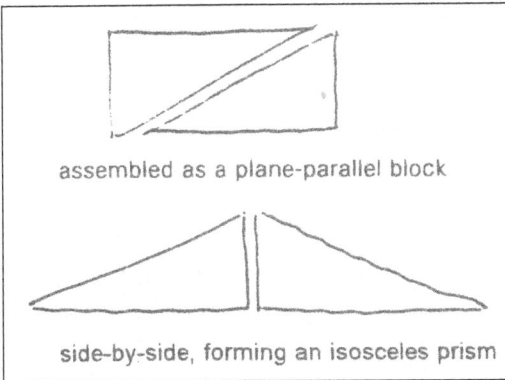

assembled as a plane-parallel block

side-by-side, forming an isosceles prism

Figure 2. Setup of a double prism (top view)

If we put two prisms together, forming one block with parallel planes (see Figure 2, upper), things seen through it are hardly displaced at all, and colors no longer arise (Op5). This corresponds to an undistorted view through a window pane of parallel surfaced glass. (The specialist/ technician can calculate the amount of refraction remaining due to the thickness of the plate. However, this depends only on the thickness of the pane, not on the distance from the glass to the viewed object; also, it doesn't take the visible surroundings and produce colors. We will disregard this plate-refraction in the current lessons.) On the other hand, if we place the two prisms together, forming one isosceles triangular double-prism (Figure 2, lower) a student standing in the center, directly behind it at a moderate distance, will appear replicated (Op6).

If the student moves closer to the prism, then in the right prism we see the right part of the student's image, in the left prism the left part, and the middle part in both prisms. As seen in this prism arrangement, the distance between his shoulders is always greatly enlarged—they are seen further apart. Through this, have we created an enlargement? No. Distances *within* each part of the image are not expanded, though a part of the image is partially doubled; we don't yet have a useful magnification. Then how can we achieve it?

Consider this double-prism once more. If the viewed student moves slowly back, while remaining on the center axis, the following sequence occurs (in bird's-eye view, left, and as seen through the prism, right):

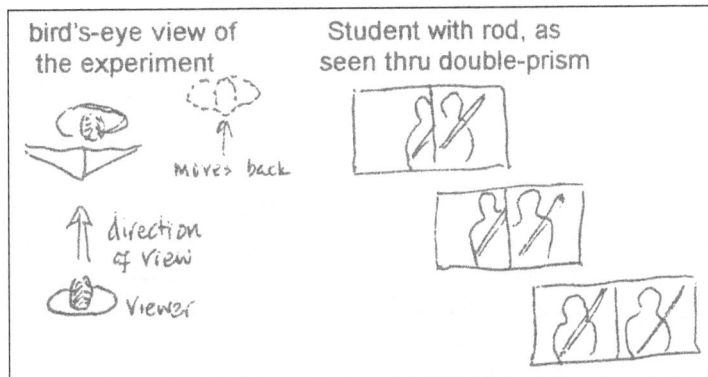

Figure 3. Displacement in two prisms

As the viewed student moves further backward, his images separate sideways out of the prisms. As his image moves rightward, his image moves leftward in the left prism. We never find true magnification, rather *only* displacement or *shifting* of the direction of something seen in the prism from its direction in actual space.

IV. ABOUT DISPLACEMENT AND MAGNIFICATION

To simplify things, think of a linear body of only five elements (a = left shoulder, b = left collar, c = throat d = right collar, e = right shoulder). This is how we would see the body without the prism:

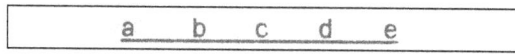

Figure 4a. Linear displacements in our prism

In the scale of the above drawing, the doubled image produced by the double-prism would be seen as follows:

The body as seen:

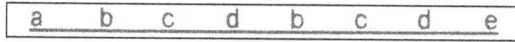

Figure 4b. Displacements in our water prism

Between the tangible and the visible a displacement has occurred. Consider for a moment only the right half of the double-prism. The visible position of each element is shifted equally far to the right of the tangible position:

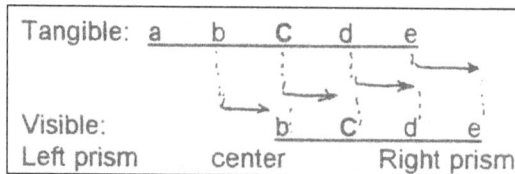

Figure 4c. Linear displacements (cont'd)

With a greater displacement, this will lead to a separation of the two images, seen at first as a double chest, later as double shoulders, etc.

In contrast, we can imagine how it would be if every element was displaced *in proportion to its distance from the center*. Initially, we are clear that the middle element (i.e., c) theoretically would maintain its position. The elements furthest outward in Figure 4d (element e) would be displaced the furthest. We obtain the following relationships:

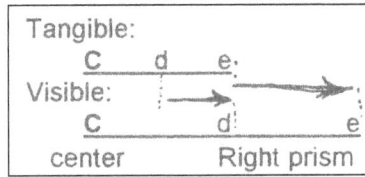

Figure 4d. Displacements creating magnification

This would be a true magnification. How can we achieve this? The displacement does not depend on the thickness of the triangular water prism which we look through (then every image in the prism would be compressed?). It depends only on the angle of the acute corner (Op3). Point d (in figure 4d above) must be viewed through a prism with a more acute face angle than point e; and, point c must be viewed through a plane parallel block (see Op5). These three prisms can be assembled into a stepped-prism (Figure 5b).

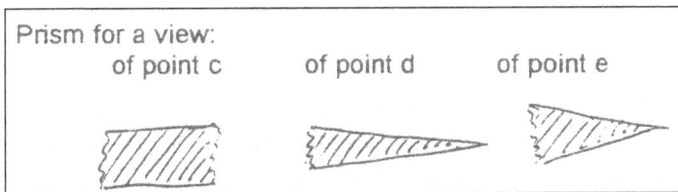

Figure 5a. Face angles on prisms for magnification

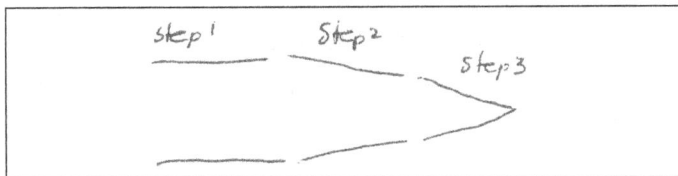

Figure 5b. Three prisms can be united into one stepped-prism

Now we are very close to a lens. If we consider further that points between c, d and e should all be displaced in proportion to their distance from the center (from c), then an infinite number of prism angles would be needed: the prism would be rounded into a lens.

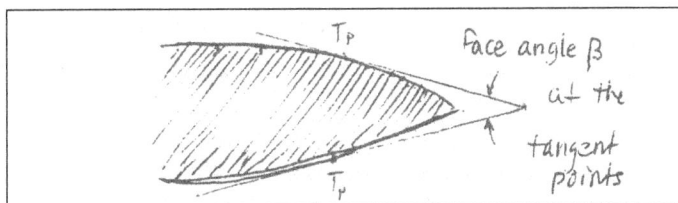

Figure 6. The rounded 'prismatic' lens
Note: The curved surfaces are not always a circular arc.

Planes tangent to two opposite points on the lens will form a prismatic angle B (interior angle). We can quickly imagine that thick, very convex lenses, will display stronger optical effects than planar, thin lenses with a small prism angle, approaching a planoparallel glass disk. We now investigate the way a large convex lens magnifies (Op7). This happens initially as we set aside the prism: First, if the seen party moves right up to the lens, it will hardly be magnified. The further they step back, the larger we will see them.

Also, we note: the greater the magnification, the smaller the field of view and the smaller connection it has with the scene surrounding the lens. To emphasize: The entire lens can be filled with just one freckle, but around the lens we see only the neck, the top of his hair, and the edges of the ears. The whole ordinary face surrounding the freckle is lost. Just when we get really good magnification, we lose orientation in relation to the surroundings. The enlarged subject is no longer characterized in a physiognomic relationship to the surroundings (like a face belonging to a scene). Therefore, we can hardly distinguish it qualitatively. Goethe says that magnification alters me, since things which I know as small are seen large. He points to the fact that everything in nature has its characteristic size, a unique expression of its place in the natural environment

V. THE SWIMMING-PLACE AND BURNING-POINT

As distance to the viewed object is increased, the magnification [as seen through the lens] increases steadily as the object moves further away. Then, a swimming or dissolving occurs at a particular distance from the lens-to-object.[2] The image looks increasingly as if it had been 'bled' by rain: broad colored bands arise as in the acute angle corner of a prism. From the edge inward, the image begins to dissolve into large, puffy, cloud-like forms, which billow out or shrink toward the center with the tiniest movement out or in (Op8, Op9a). A sort of chaos arises as the so-called *swimming-place* is reached. For a very distant observer, this swimming occurs when the distance to the object equals the *burning point* (manufacturer's conventional focal length for that lens). If the object (or a newspaper backdrop) is moved further beyond the chaos of the swimming-place, a large image now re-appears, but upside down. It then grows smaller and clearer as we shift the object still further away.

So, for an object *just at* the swimming-place, the blurred appearance is as if it was infinitely magnified. Also, the object appears spread out to the periphery of the lens, and we see an infinitely small speck of its image enlarged to immense size, usually a formless colored surface. We can complete something like the sketch in Figures 4c, d.

Moved further back, it is as if the view of the object went 'beyond infinity' to come back from the other side, and therefore inverted. This is purely a lens-effect not seen with prisms.

However, this is not the time to go deeper into how prismatic refraction is continued over to the other hemisphere of the lens, is collected into the center, becoming magnification in such a way that we see one little spot spread out over the whole lens surface, presenting an infinitely enlarged and widened world.

Without investigating the inverted image further, however, we do want to go into other possibilities of magnification without inversion—the loupe or burning lens.

Using a sufficiently large convex lens, each student should measure the enlargement (Op9). Simultaneously, the loupe makes the eye very shortsighted. The maximum distance we can see objects (enlarged) extends only to the swimming-place—not beyond.

With glasses, things appear closer when we place the convex lenses directly in front of our eyes; we see the object at a reduced distance and at an enlarged size. Simultaneously we can notice how the eye is completely relaxed, as it is when viewing the distances in the normal way. If we want to treat a far-sighted eye (which can look at distant objects only under tension—youthful far-sightedness) and aid it in seeing nearby objects, we utilize a convex lens which makes it just a bit shorter sighted. The eye can then see close things in focus with less tension. In contrast is the near-sighted eye, which has to bring everything close by as if it cannot relax enough; if we want to help such a person see their more distant surroundings clearly, we have to provide outer relaxation by taking away the excessive convexity of their eye lens. Therefore, we use a concave lens.

The **burning distance** is the distance of an object seen in chaos (as described above) by a far distant observer (theoretically, infinitely distant) (Op8); objects at 1 burning-distance (1 BD) will appear like this only with my eye far away. Inversely, at this distance, chaos would also be seen *from the viewpoint of that object*. In other words, if I put my eye at the swimming-place, an infinitely distant scene will appear infinitely expanded, filling the lens with a featureless or cloudy expanse. So, if I were to look at the sun (behind the clouds, as direct sun would be very damaging!), the sun disc would fill the lens. For a paper surface, the sun normally occupies only ½ a degree in its cupola. If I place the paper where my eye was, the sun appears larger, filling the entirety of the lens. So, it acts proportionally hotter—even burning. Thus this distance (swimming-place), where such a distant light fills the whole lens, is the lens's burning distance (focal length). In other words, the swimming-place for an infinitely distant observer is equal to the focal distance of the lens.

VI. STEPS VIA SEVEN-MILE BOOTS TO A TELESCOPE

It is not the task of grade 8 optics to present a complete chapter on optics. We aren't able to follow an unbroken thought-sequence to the telescope—as we did from prism to lens. So, once the students have applied these ideas about the basics of lenses to the metamorphosis of images, many other ideas can be presented in connection with them, more as an overview from the point we have reached; however, optics is soon ended.

For example, as before, we examine the magnifying properties of a lens, not only how it creates new images (enlarged and inverted), but also the way it creates *free (projected or 'real') images* on a piece of paper. With the pinhole camera (Op11), we have already met such free-standing images; they arise on any kind of whitish surface independent of an eye, and they are visible to many people simultaneously. Since the viewer doesn't look through or at the lens, but looks from any direction toward an optically neutral place in space (the so-called screen), he hardly knows if the image is a painting done previously. Here, we are no longer dealing with a transformed view of an object toward which we look with the aid of something (a water layer, a mirror, a spherical glass body). Rather, we are dealing with an image which magically appears on a surface, free in space, and can be viewed as if it was any other object in space—the view of the object is detached from its source, and therefore such images are endowed with technical power.

As experiment Op11 shows, it is not possible to make a telescope with one lens. The major discovery of looking through two appropriately sized lenses, was probably made by an unknown glassworker around 1600. How the astronomers Galileo Galilei and Johannes Kepler took it up and made the *telescope* (far-seeing device) is well-known. We will construct the so-called Keplerian or astronomical telescope, consisting of two focusing (convex) lenses. Since the viewed scenery unfortunately appears upside down, we cannot avoid introducing the newer concave lens objective used in the Galileo's telescope (Op12).

Fundamentally, how does the telescope make distant objects appear closer? The lens oriented to the world of things we view (objects) is called the objective. This objective can form a 'focal image' at its focal plane, as we saw in Op11. The larger this focal length is, the larger this image must be.

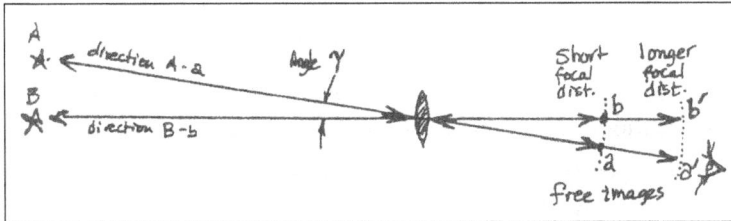

Figure 7. Angular magnification by oculars with various focal lengths

For example, from point a in the focal plane, a person can view object A in direction a-A through the center of the lens, which is essentially plane parallel there. Object A could be a star infinitely far away; the distance is immaterial, it simply lies in this direction. Object B lies in another direction. With this we don't mean that an image has to appear between a and b on a sheet of paper at the focal plane; rather we are only showing how it is that, when these relations prevail, we get the distance a-b. (Naturally, this doesn't come to expression by virtue of a 'ray' through the lens center to direction A, but because a person looking outward from a sees brightness and color relating to A over the entire lens). If the lens had a larger focal length, then the larger distance a-b′ would occur. The size of the free-standing image formed by the objective is directly proportional to the focal length of the objective lens. However, nothing better is achieved if we make the objective essentially flat in trying to give it increasing focal lengths, to achieve greater enlargements. Such images certainly would be very large, but they would also be very pale and washed out. If telescopes or field glasses are high power (enlarging), they must therefore have not only objective lenses which are rather flat (although not totally), but also quite large; still, this too has limits in practice; from the diameter of the front lens, we can tell if the binoculars are really usable as night glasses or not.

The second lens lies nearer the eye and is called the ocular (or eyepiece). It is inserted as a small 'loupe' or magnifier to make the 'free image' appear larger; it must be placed close to the focal distance of the objective. The more convex the ocular (i.e., the smaller its focal length), the larger the magnification. The enlargement is inversely proportional to its focal length. (See supplemental topics such as enlargement with a magnifying glass.) The magnification of a telescope increases when the focal length of the ocular decreases. So, the magnification M depends on the focal lengths of both lenses. Put into a formula, *magnification* M is given by:

$$M = \frac{(f \text{ Objective})}{(f \text{ Ocular})} \qquad \text{where } f = \text{focal lengths}$$

To be precise, this enlargement has to do with angular size not with planar size. If we see the moon with naked eye as ½° large, we would see it in this telescope as M x 0.5° large. The whole telescope optics are equal to the sum of both focal lengths, and its tube must be at least this long.

In sum, we have explained almost everything about visible images and their transformations—without using much geometry—and, we have even given a plausible clarification of the formula for telescope magnification. Using this equation, any student who wants to could build a telescope, e.g., with a 20-power magnification. The actual problem lies not in determining what lenses to use (the ocular is glued in the end of a sliding cardboard tube) but in achieving a vibration-free tripod support. Once this is accomplished it is possible to view the moons of Jupiter as Galileo did—not in order to create a celestial science based on telescope observation, but rather *to allow the students to experience how a person can peer out into the far expanses of the world and meet new things.*

Experiments in Optics

OP 1 LIFTING (REFRACTION): INTRODUCTION TO COLORED FRINGES

Using an aquarium or large glass tank, at least 70-80 cm (27-31") long:

a. Cover the inside of the glass tank with stiff, waterproof paper, white on the bottom, dark on the sides. Mark the paper with a black strip, made with magic marker, 2 cm (¾") wide, for example. Initially lay or place a few white tiles against the narrow wall, with horizontal black strips of adhesive tape. Fill the aquarium with sufficient water, which can be removed by the students as needed. The design on the end wall will be squeezed upward, and the designs on the bottom stretched out, all lifted upward toward the far, narrow end wall.

Simultaneously, many colored bands appear: a blue-violet, bleeding up over the lower edge of the black strip; a red-yellow band at the upper edge of the black strip (further from the observer). We could make these things clearer with two almost full tanks, one of which is then filled to the brim with an extra pitcher of water. It would even be better to use tin water troughs, as they have no distracting reflections.

b. Move around next to the long wall and observe the seam where the bottom meets the opposite wall. It appears curved, to sag down and away to either side, most strongly in front of us. Notice that even the narrow wall joint has a slight curvature.

c. A penny is laid flat on the bottom; another is held outside and raised until it appears the same size as the submerged one (when seen perpendicularly). The outer one will be positioned at about 25% of the water's depth. This can also be demonstrated using a graduated cylinder filled with carbon disulfide, with only about 20% lifting (see Table on p. 32).

OP 2 GOETHE'S ARCHETYPES: LIGHT AND DARKNESS

Fill a smaller glass tank (about 10 liters, 2.6 gal) with water and set it on the window ledge. Add a few drops of liquid soap or milk.

a. Put a black card behind the tank and a light to its side. In the very dilute, milky liquid (looking through the side-lit water into darkness), we see a whitish-blue. It is stronger the more dilute and more intense the side-light. An especially thin milkiness is called a 'blue haze' as an expression of an insubstantial, immaterial, illusory phenomena.

b. Cover the top and sides of the tank with a paper tent. A yellow-brown tint can now be seen by the students looking through the milkiness, dimmed by the surrounding screen, into the brightness beyond. It is stronger the denser the 'murkiness' of the water and the dimmer the back-light. The more the turbid media is evaporated and concentrated, the less of the bluish tinge can be seen, since it is now darker itself and less able to be lightened. This now too-dense turbidity produces an excellent orange and red.

Goethe sums up all the permutations of these phenomena in two archetypal phenomena: light overcoming darkness produces the bluish-cool color bands; darkness conquering light produces the reddish-warm colors.

OP 3 WATER-PRISM DEMONSTRATION: REFRACTION

Owing to their similarity with what we have just seen in the water tank, water-filled prisms are quite useful. We have designed a large vertical prism which will allow the class to see the whole upper body of a student behind it.

a. Have a student go behind such a large water-prism and walk to and fro. No matter where he stands, the class will always see his image shifted toward the narrow edge (the acute corner) of the prism. (see Figure 1, p. 34)

b. Now rotate the prism so that the image appears in a larger corner (e.g., 60°), and we see that the displacement has increased (through the prism compared with no prism)—and the colored edges are more spread out. (Be careful, for at a very oblique view-angle, we will also get reflections, which are to be ignored here.)

c. Observe a black circle on a white card behind the prism. Notice how once again colored bands are seen, and the design appears stretched or elongated—but now laterally, left and right! (i.e., always toward the edges of the prism, or perpendicular to its axis, as in Op1).

OP 4 CLASS EXPERIMENT: GLASS PRISM

After the class as a whole has seen the horizontal refraction in the large water prism, it is not difficult for the students to understand refraction in all sorts of applications. We begin with a single glass prism, initially oriented to looking through the acute corner, and then we introduce the double prism. Discarded field-glass (binocular) prisms frequently can be substituted for a double prism.

OP 5 PLANE-PARALLEL BLOCK PRISM

As described in the main text, build on the simple observation of Op3b to show how, as the face angle gets more and more acute, approaching 0° (a parallel block, like with two right-angle prisms placed hypotenuse-face to hypotenuse-face), eventually no shift is seen. Then show how an increasing face angle corresponds to an increasing refractive shift. (see Figure 6)

OP 6 DOUBLE-PRISM

As described in the main text, place two right-angle prisms together at the small face, forming one broad isosceles triangular prism. Note how the two split halves of the image are shifted apart (a rudimentary magnification). [You can use two small water prisms stacked, and view a magazine page.]

OP 7 MAGNIFYING LENSES (OBJECTIVE)

Fasten a large convex (or focusing) lens, e.g., with f = 50 cm (20") and 12 cm (4.5") diameter (Leybold) at head height When a student puts his face immediately behind it, essentially no magnification is observed by students at the other end of the room. If he goes 20–30 cm (8–12") back, enlargement increases; parts of his face are no longer visible through the window of the lens, without reappearing around it. This may be sufficient for a comparison with the spread-out images of the double prism.

In introducing the lens, it is important to point out the curvature. Have a student explore the various thicknesses of the lens at the edge, in the center, and near the opposite edge. Then, draw a large cross-section diagram of the lens on the blackboard with the fingers touching and exploring the thickness.

OP 8 SWIMMING-PLACE

This experiment is done similarly to Op7, but have the students stand together in the rear of the classroom so they see all the chaos at the same time—that is, at the same distance of the viewed object from the lens. To show the inversion (180° rotation) of the image in addition, a large printed paper could be used instead of faces behind the lens. We will find the same distance to the swimming-distance where the image dissolves into chaos, just before inversion as we did before. This may be approximately 1 cm more than the stated focal length of the lens (within the accuracy of these measurements) due to the observers' not being infinitely far away.

OP 9 CLASS EXPERIMENT WITH A SMALL MAGNIFYING GLASS (LOUPE)

Provide half the class with a thick convex lens with as large a diameter as possible, e.g., 5–6 cm (2–2½") dia. and 6–8 cm (2¼–3") focal length:

a. Lay the lens on graph paper. Lay a second piece of graph paper over half of the lens. Examine whether the grid seen through the lens is enlarged (a tiny bit), and whether a bit of the underlying graph paper is not visible. [Yes, a small bit has to be, due to enlargement.]

b. While one of the students stands on a stool, have the others slowly raise the lens above the paper. Measure the "blurry-distance" as the distance from paper to the middle plane of the lens when the viewer on the stool sees the image dissolve into chaos. [It will be essentially equal to the "burning distance," as measured in Op10.]

c. By comparing more graph-paper diamonds as in Op9a, determine the magnification when the lens is at: ¼, ½, ¾ and ⅞ of the swimming-distance [about 1.2, 1.5, 2.5 times as big].

d. Graph-paper lines move with an even speed when moved toward and away at distances between paper and focal distance. When does the image swell faster? [Near the focal distance]

e. Sketch the bulging pillow-form seen near the edge of the lens for a object at a distance close to ¾ of the focal-distance. Paint the colored fringes seen there. Connect this observation with Goethe's prismatic color archetypal phenomena.

f. Observe and then sketch the image seen outside the window when we look through the lens held at arm's length. [smaller, inverted]

g. Examine how many cm distant your lens must be held away from your face so that something seen outside the window (or, if necessary, the window frame cross) swims in total chaos; this is the burning point [the focal distance].

h. Observe a long, delicate pencil line with a magnifying glass. Note the individual graphite flakes and the paper fibers, the ridges and folds of skin, etc. What we see in this way can be seen by straining the naked eye, but this is at the limit of 'direct phenomena.'

OP 10 BURNING-DISTANCE (FOCAL DISTANCE)

Place the large lens of Op7 & Op8 in a sunbeam or the light from a very bright lamp in the corner of the room. Now move a paper on a cardboard back and forth, showing how the smallest spot arising at the swimming-distance has the greatest brightness. It will happen often that the paper starts to char at this distance: This is the sun, imaged by the lens.

OP 11 CLASS EXPERIMENT: FREE IMAGE ('REAL' IMAGE)

Mount the smaller lens we studied in Op9 so it faces a window. Now (with the room lights dimmed), shift a paper 'screen' closer and further behind the lens. Suddenly, the window frame stands out sharply and somewhat nearer the lens, a landscape scene now becomes visible. The distance from view screen to lens is about equal to (usually a tiny bit larger than) the focal-distance. Compare the clarity of the card image with a direct view of the landscape—the landscape has been brought closer! But a magnified image of distant things is not possible with a single lens, only bigger images of nearby things (and correction of near-sightedness, as could be experimentally shown with lenses of various focal lengths and near-sighted students).

OP 12 TELESCOPE

First demonstrate focal distances of the large 'objective' lens from Op7 [with approx. f = 50 cm (20") focal length and 12 cm (5") diameter], and also of a second convex 'ocular' [with about f = 5 cm (2") and 4 cm (1-9/16") diameter], as in Op10. Then, assemble the two into a telescope. Mount them about 50+12 = 62 cm (24.5") apart; we get about 1°x magnification. Keep the centers of the lenses exactly aligned, using a board which can be elevated or lowered at one end to point it at various landscape scenes about 300m (984') distant—if possible, including a pine tree top, with other branches about 30m (98') further in the distance.

Have a student place his eye near the ocular in the center of the lens; a clear but inverted image will be seen. When looking just a little bit diagonally, colored fringes or even fuzzy clouds will fill the image, because this optical system has not been corrected for optical flaws, especially chromatic aberration.

To view closer objects, the two lenses must be separated slightly (pull the ocular toward the viewer). Now the 'free' image of the closer objects created by the objective lies just beyond the focal distance of the ocular (a bit further from the objective), so the magnifier for this image (the ocular) must be pulled back to 'catch' this relocated image. It is very nice if we have two of these telescopes and the students can look through one after the other to establish that:

a. the distant pine tree's tip, seen in the telescope, shows more detail than is possible with our naked eye, as small birds and even distant airplanes can also be discerned.

b. for closer objects, the ocular must be pulled back (closer to the viewer).

Meanwhile, the others can work on an accurate scale drawing of the telescope construction in side view, showing lens curvature, cross-section, and distances; and, in a second view, showing the tree top as seen without and with the telescope.

Grade 8 Heat

The other topics for physics in 8th grade cover a lot of ground. The study of heat is carried further in grade 9 and may therefore be a bit shorter in grade 8. The themes could be formulated as follows: convection, radiant heat, heat technology.

I. CONVECTION - RADIANT HEAT

Using a high-form beaker or the glass convection loop apparatus (He1), we can show the streaming movements which occur when we heat the center of the bottom intensely. The movement and streaming of the flame appear to be carried over into the movement seen in the water, as made visible by dye or paper fibers in the water. Then, heat the convection U-tube and note the flow effects of heating; note the unequal water level at first, then the water *movement* as the 'bridge' is inserted. Before this, the two water columns were in equilibrium, even though the heights were different; now, with a path provided, the disequilibrium of height can be transformed into circular movement.[1] If we take away the heat source, the movement will finally cease once a new equilibrium has been attained. In summary, we can show convection with the usual teaching device of the 'convection tube.' As soon as heating is begun, a circulation appears.

It is very important to not merely portray convectional movement as an upwelling of something warmer; initially, we can see the driving force is colder and warmer water *adjacent along a broad interface*. The cold water presses down to the same extent as the hot water presses up. To this extent convection is a phenomenon of pressure (hydrostatics) or aeromechanics, treated—usually later—in a general way in grade 8 where water- and air-pressure phenomena are given an exact study. Air convection can now be shown using smoke blown around some sweaty students or some cold objects. This could also be shown using a Japanese paper lantern, heated inside with a candle, or else with hot-air balloons or a candle-powered Christmas carousel. The same sort of warmth relationships are seen in the movement of air and water on earth. Things found in the laboratory should not be presented as the cause of these natural phenomena, but rather seen as a micro-image of them.

We can fruitfully examine: water stratification in lakes, formation of lake-ice in whiter, etc. (see heat lore, Grade 7 book) Thermal stratification of water (thermal inversion) can be shown with the slanted beaker experiment (He3). However, the thermal layers which can form in air are best treated with the rest of aeromechanics and meteorology. Technical applications include space heating via radiators, which should be placed at the outer walls where the air is cooler, denser and sinks; there we can best create an even mixture of the rising warmed air from the radiator with the sinking cool air. (If, however, we maintain all spaces at the same temperature—room, patio, auto, office—our warmth organism does not have to exercise itself and many colds are passed around).

II. RADIANT HEAT

The archetypal image of radiant heat is the free passage of heat through empty space, e.g., connecting the earth's surface exposed to the dark night sky of a winter's night or to the bright sunlit sky of a summer's morning. Warmth does not come "marching toward" the earth, nor cold away from it, nor does it "ray out" of the earth. Rather, in open space, a change in warmth occurs at the surface of the earth. Down here around us, warmth is called forth from the outermost periphery of our surroundings, without any phenomena of being transmitted by a physical medium. The supposition of some sort of passage of energy along myriad 'rays' interposed in space sidesteps the phenomena; it fashions concepts which certainly can grow into technical apparatus, but do not form a basis for a unified understanding of the *gesture* of how heat arises. And, as is often the case with technical apparatus, we have to leave open the question whether other concepts, such as those we have striven for here, couldn't also beget the technology we need (whatever we deem necessary).

Concretely, this orientation of the earth to the periphery actually involves only the uppermost layer, only the top millimeter of the earth's surface which presents itself as *a kind of earth warmth-organ for the conditions of the surroundings*. A sunlit patch of earth fluctuates by as much as 60°C between sunrise (minimum) and one hour after high noon (temperature maximum); all other sites deeper in the earth do not vary so much. If one of these sunlit patches is air-covered, then warming occurs; if meters of water or ice cover it, then little or none. Inside a glacier ice-cave, even when the sun shines through the ice, it will not get warm; a cold illumination spreads out inside there and only out on the exterior, air-covered surface of the ice will a slight warming occur. A fish which swims into a sunlit space in the water will feel only the barest trace of warming (the swarms of small fry therefore swim usually in the upper layers). Also, when the sun shines on the bottom, it hardly gets warm there; rather it is the uppermost layer of water which often gets warmer.

The zone only a hand's breadth above the earth's surface is the organ of the earth for the warmth of the surroundings, for the illuminated surroundings it is more the water bottom (with clear water), Should we now describe these phenomena by the concept of radiant heating passing through the air but not through water? The fact is that only objects in the sun—as well as being simultaneously illuminated— generally get warmer, as though they were *summoned to heat*.

A person should not conceive of warming as a condition traveling through space, the encounter with which would be the cause of the phenomena (seeming to be fundamentally an expression of raying and conduction). Rather, we can express it thus, and remain entirely in the phenomena: Heating means a body *accompanies* or in general it *adapts* or adjusts to the surrounding warmth and coldness. "Summoned," "accompanying," "adjusting"—nothing depends on the words themselves, but *on appealing to actual experience*, in order to be able to later link up to the all-encompassing connections. The inclination to explain what we experience as a product of unexperienced processes in space implicitly assumes nature 'dives down under' the perceptible world, to become active invisibly (as if the heat rays, in some other situation, could be visible to us—which is not actually possible!). This inclination derives from the decision to consider the mathematical-kinetic theory as the real nature of heat (Kant and logical positivism).[2] This must not happen in our classes.

If we try to renounce 'rays,' in order to be free to be active with our thinking *in the phenomena*, then we also renounce the concept, the mental model of rays. However, we don't need to renounce the word *rays* or *radiation*. On the other hand, the students should know that people can think in that way and know that, with such concepts of rays, the situation is not wholly explained just by bringing something into our grasp through quantified measurements or calculations (although they may enable us, for example, to construct infrared night glasses for explorers).

What do we *actually* find connected with these so-called 'ray' phenomena? An open, free bestowal of heating, which, however, *does not engender any heat phenomena in the intervening space*. Eventually, we study the process of radiant heating using a blackened mercury thermometer and an air bulb thermoscope

(He4). Radiant heating is diminished when it involves anything other than open air. Also, heating happens in the dark—no (visible) illumination is required for it. The surface characteristics of the hot object and the thermoscope are important, too. (see He4)

With these demonstrations, we could express it thus: A body which causes another to become warm or cool through open space, must itself be warmer or cooler than the other. The one body **takes the other along with it**, causes it to assimilate, to equalize to it. For the body assimilated, it is a question of pursuing warmth or cold. The free space can be quite extensive; the assimilating is governed by the same rule we found with illumination: It depends on the size of the illuminating dome (see 'cupola' in Grade 7), i.e., on its angular extent in space. Usually, **uptake of warmth and uptake of illumination occur simultaneously**—the bright sun brings warmth to my skin. However, we can separate the warmth (or cold) connections from the brightness-connections (winter night or experiments with 'dark heat'). We work through the experiential differences instead of the outward similarities.

Generally, (1) relationships in brightness are most striking at the surface, (2) warmth relationships develop when we penetrate deeply within materials, and (3) cold relationships take place in the dark. It is not easy for the conventional natural scientist to take up the foundational ideas of these distinctions. But, if he manages to do it, then he could work out in more detail the similarity with static light in technology: to introduce an aluminum parabolic (concave) mirror as a connecting link in the orientation of a surface to the glowing wire of a heat lamp; here the line of the filament appears enlarged, as seen from the illuminated surface, (i.e., instead of just a tiny glowing wire, the whole disk of the mirror appears bright). With such mirrors, one can (re)direct the brightness connection (as in stage spotlights).

III. HEAT TECHNOLOGY

Three technical devices may be introduced here: the thermos flask, the solar heater, and the bi-metal thermostat. The first two can be found described in many conventional physics texts. For the bi-metal thermostat demonstration, use a large strip: 4 x 15 cm (1½"-6"). Show how it responds to heating or cooling by flexing one way or the other. Perhaps the different expansion coefficients of the two metal layers sandwiched together in the strip could be mentioned, if the question is asked.

IV. NATURAL IMAGE(S)

The kind of warmth connections we have studied so far can be arranged by their relation to the surface of the earth. We can think of these terms as technical abbreviations for the whole natural processes: **Conduction** describes the metamorphoses of warmth [heat transformations] which we imagine to prevail if we penetrate deep into the body of the earth. **Convection** expresses our conception of how we experience the interchange of warmth in water, although conduction is active to a degree in water, since it is more substantial. In the atmosphere, convection is decisive. In the air about the earth's surface (and on water), all three processes come together, but **radiation** prevails, invisibly linking one region with another.

Moreover, in addition to these three nature processes, and after evaporative cooling as a further topic, comes heating of air by condensation and compression (expansive or adiabatic cooling). These are a modern formulation of scientific ideas. A full treatment of many of these topics in the physics of gases is deferred until grade 9.

Experiments in Heat

HE 1A CONVECTION IN TALL-FORM BEAKER

Fill a 2-liter, high-form beaker, with water and place it on a wire mesh over a burner. Just before lighting the burner, place 3 tiny grains of potassium permanganate in the center of the bottom. [TAKE CARE: Do not touch the crystal, as its chemical stains are difficult to remove.] Or better, use colored dye, a few drops of bluish color. The strong color shows the streaming forms that arise at the bottom, as it is heated over a Bunsen burner. Note the direction of streaming, the shape of the currents and stream lines, the intensity in various locations nearer and further from the heated area.

HE 1B CONVECTION CELLS

Mix a small quantity of fine aluminum powder into a somewhat viscous oil (e.g., linseed or motor oil). The fine particles will show a network of darker lines marking off little 'cells' of convection. This also occurs in a cup of hot coffee, when cream is added slowly; for a brief while the miky coffee will begin to mix and show the convection cells.

This demonstrates a fundamental phenomenon of heat: bringing everything into movement. In its highest manifestation, warmth of heart, it also reveals this quality: opening and extending the soul-warmth of a person beyond the boundaries of the person and moving the deepest sheaths of the soul.

HE 2A CONVECTION CURRENTS

Then, heat a U-tube apparatus (see below) at one corner. Again dye the water with a grain of permanganate tapped gently so it falls into the center of the U-tube. Or introduce a few drops of coloring as described above. Note how the water stands higher at the heated arm of the tube.

Carefully insert the water-filled siphon bridge until the ends are in the U-tube arms and submerged. Immediately after the bridge is in place, the U-tube water level begins to equalize. The aperture/outlet of the siphon-bridge which is submerged in the warm water column experiences a greater water pressure than the opening on cold side, and the warm column stands higher. The cold water column thus receives an influx from the bridge. So, the warm water column now flows upward, while the cool (and initially shorter) water column sinks down and slides across the bottom toward the heated corner.

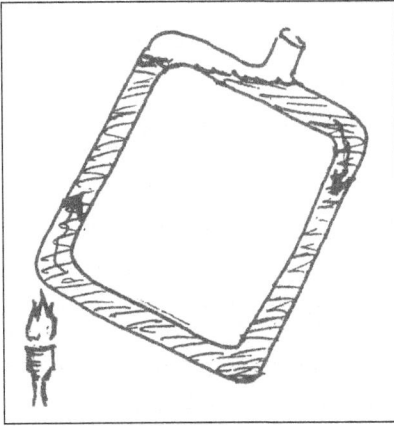

The flow stops slowly when the burner is removed. The hot water ends up floating at the upper part of the bridge, the cool water having sunk to the horizontal tube at the bottom. Equilibrium now returns; motion ceases.

The demonstration can be repeated with the standard, one-piece convection loop-tube demonstration apparatus.

Figure 8. U-tube convection demonstration (with siphon bridge)

HE 2B AIR CONVECTION

Demonstrate air convection caused by warm bodies by collecting several sweaty students in a group in a corner (perhaps after they have run up and down the stairs quickly). The observer should be positioned a bit removed and the heating turned off. Blow smoke (made with pipe tobacco, for example) around their heads; note how it rises up. The observers should be kept back, so as not to disturb the experiment. Compare with another group of cold objects (boxes from the school kitchen freezer) by blowing smoke around them; see how this smoke flows downward.

Alternatively, heat a very thin rice-paper balloon, closed at the top, with several candles inside, sticking up into the bottom hole. Allow it to begin to rise (it may even begin to burn, and rise into the air as it burns up). Discuss hot air ballooning.

HE 3A THERMAL STRATIFICATION

The slanted beaker experiment shows thermal stratification quite well. Hold the bottom of a high-form beaker in your bare hand, inclined about 30° from the vertical, and allow the flame from a burner to heat just the upper 2 cm of water until it boils. TAKE CARE not to splash the water up onto the hot wall of the beaker; keep the flame on the water-covered glass wall only, not above. If this is done properly, there is no difficulty holding a beaker of boiling water in your bare hand. You may want to practice this with an insulating glove until you to get the knack.

HE 3B THERMAL MIXING

Use two similarly shaped jars. Fill one with cool water dyed blue, and the other with hot water dyed red. Place a square of thick paper over the mouth of the hot one, and invert it over the mouth of the cool one; the colors don't mix. But, when we invert the whole apparatus, with the hot on the top and cool on the bottom, the colors readily mix, showing the convection currents in the water.

HE 4 RADIANT HEATING

Prepare a red-hot iron bar. Assemble: 1) an air-bulb thermoscope (see grade 7 Heat), from an Erlenmeyer flask, sooted on the outside, having a U-tube water trap in the cork to show air expansion; 2) a 'radiant thermometer,' from a mercury thermometer with a sooted copper foil 'flag,' shaped into a flat-plate collector around the mercury bulb. Show how each responds to a red-hot iron bar held a few cm away from both (the air shifts the water up; the mercury climbs).

Now, observe how they behave when held behind a double-paned glass wall, with continuously refreshed water in between, and exposed to nearly-glowing fire bricks (constantly replaced as they cool)—much weaker warming. This is done in a darkened room to show that visible light is not necessary.

Show the difference in warming of an air-bulb thermometer exposed to a red-hot iron bar, when variously coated with (1) flat black soot, (2) flat white (by the smoke from burning magnesium underneath beforehand—CARE for your eyes!), or using (3) shiny aluminum foil. If we place a beaker of boiling water nearby (rather than the hot iron bar), it makes a difference whether the beaker is coated with flat black, white or aluminum. The variations in visible connection to the surroundings (how bright or dark something appears) can be shown in a fascinating way by heating to glowing heat: (1) silver with a sooty stripe, or (2) carbon with a chalk stripe. In (1), the otherwise dark, sooty carbon surface region glows quickly; in (2), the bright, chalky region doesn't glow as easily.

HE 5 HEAT TECHNOLOGY: THERMOS; SOLAR PANELS; BIMETAL THERMOSTAT

a. Show how the various types of thermos bottles will greatly reduce the adaption of cold contents to the warmer outside. Demonstrate the differences for: evacuated double-glass walled flask (1) without silvering, (2) with silvering, (3) without a vacuum between the glass walls, (4) with and without silvering. [Evacuated and silvered is best insulator; just silvered is also quite good, since they both nearly eliminate radiative heat exchange.]

b. Discuss solar water heating panels; various kinds of reflector materials; various heat-collecting water tubes and coatings.

c. Demonstrate the application of a bi-metal strip (usually invar). Show how its bending up when heated (test beforehand which flat side to hold up) keeps it flexed away from the Bunsen flame; and if the flame is blown out, how the strip cools and sort of closes off the gas, preventing a dangerous leak.

Alternatively, the strip could be mounted inverted. Now, it bends down when heated and could control (shut off) the very system heating it, say by bending down to cover the Bunsen burner and reducing the flame, allowing the strip to cool, which then causes it to flex upward, allowing the flame to get larger again. This is a crude example of a mechanical *feedback control loop*, a system which governs itself.[3] The implementation of this phenomena in a thermostat should now be clear.

Grade 8 Electricity

In grade 8, electricity and magnetism can be brought together; and, since electricity and magnetism are not studied separately, this part of the physics block begins with a fundamental understanding of electromagnetism. Developing a basic understanding of electromagnets leads to a grasp of technical devices which permeate our modern civilization in countless numbers, as well as in every facet of our daily life: relays and motors, measuring devices (dials & gauges), bells and—significant historically—the Morse apparatus. **The technical products of the modern world surrounding the students are actually *human thoughts* which have become implemented and now meet us from outside**. The thoughts should be penetrated by the students, so they can find their way out into the world. The electromagnet (as a copper coil with an iron core) is never found outside in the free activity of Nature. Initially, it is only brought into existence by people.

A universal usefulness and power for technical apparatus are characteristic of electromagnetism. Electricity can take over the function of almost any apparatus, because *it possesses no domain of its own in Nature*—as optics does with brightness or acoustics with sound and timbre. Electric phenomena can at one moment be expressed as light, at another as sound, at yet another as mechanical forces. If electricity formed its own domain in Nature, then it would have boundaries within it. In this **global utility** of electrical and electromagnetic effects, we also glimpse the **homelessness** of these forces, which stand at our service in such a variety of ways.

I. THE ELECTROMAGNET

A point of departure for a systematic study of electricity is Øersted's compass experiment (El1) circa 1820. It was already known at that time that a lightning strike to a ship's mast would demagnetize a nearby compass needle, making it point South! People didn't know what happened, but the effect could not be denied or reversed. If we touch the compass to one pole of a Voltaic column (battery) and connect it into a current circuit, there is no magnetic effect at the terminal. However, if (as Øersted did by accident) we place it just *beneath* a North-South conductor which is influenced by a strong current, the magnetized compass needle will temporarily lose its North-pointing ability and orient itself more or less diagonally (East-West) to the wire.

If we place the compass needle *over* the wire, it turns the other way; again it orients itself approximately transverse, but only in the area nearby (Figure 9a). Instead of a [copper] wire, we could use any metal rod (insulated or not) or even a glass tube with a salt solution [i.e., any conductor]. The transverse placement of the needle is greater the nearer it is to the current-carrying conductor and the greater the current. Also, any salt solution between electrodes (e.g., the solution between the copper and zinc plates of a voltaic cell) can also deflect a compass—if the current is strong enough.

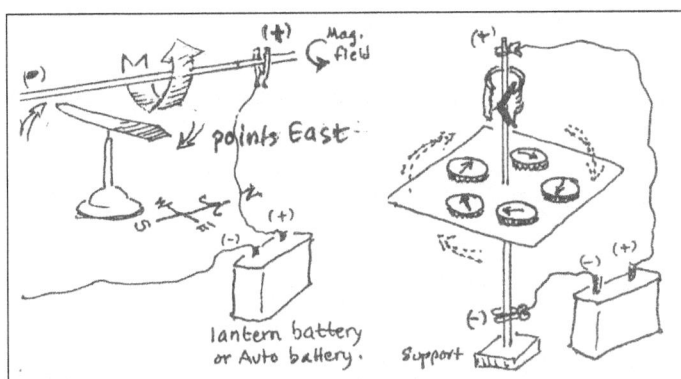

Figure 9a,b. Magnetic field surrounding electric conductors

By placing individual compasses in several locations, we can see how the various compass needles, positioned around a vertical wire (see El2) will form a circular pattern surrounding the wire when the current is switched on. (Figure 9b)

To the two phenomena of current already investigated in grade 7 (**heating in the wire, anode consumption** in the battery), we now add a third: the **circular magnetic "field"** surrounding the whole conductor path of a completed circuit. This magnetic field is plainly not a perceptible phenomenon, as we might think from the way we often speak of it, as some sort of objective 'thing.' Much more, the 'field' is a *concept* of the possible activity of a force on the compass needle, now existing everywhere around the wire.[1] We could also say: It is an *idea* of the *direction* to which the needle *would potentially point*. The magnetic force, which appears with the current, increases with current strength (measured in amperes, a concept from grade 7) and decreases with distance from the conductor. It quickly becomes very tiny, but doesn't reach a boundary; so, isn't *current* actually a phenomena of the air-filled *surroundings*, rather than of the *conductor's interior*?

In contrast to the transient electrostatics studied in grade 6, the simple current circuit—as we saw already in grade 7—closes itself off, forming a stable complex of current phenomena which depend only on the circuit itself (the current and the resistance), and not on the surroundings, as with static electricity (type of insulator, how it is rubbed, humidity, etc). Unlike a charged wire, we can even slip a circuit into our pocket (flashlight)! Now, in grade 8 we also discover that in its magnetic activity, a circuit forms a tiny earth, as it were. If a completely mobile magnetic needle is moved all around the axis of a current-carrying coil, it will describe a pattern as if it was moved from the North to South geographic poles of the earth (see El5).

Much of the material presented thus far is background information for the teacher. Yet, the students will understand the following amusing assignment: a disguised pirate living on a boat. He is traveling North (e.g., from the Curaçao to Hispaniola) on a starless night. With a wire concealed underneath the compass, he attempts to divert the ship off course, perhaps to a pirate island to starboard (West). Should he place the wire across the beam of the ship (E-W) or along it (N-S)? [parallel to the beam, i.e., N-S, and with the North end the positive terminal]. In practice, when the deflected compass remains 'stuck' to a NW point on the compass rose, even when the ship has shifted course or begins to circle, the captain will certainly head for the nearest island. The following is also an interesting question: Can we tell if an E-W oriented wire is under the influence of a current solely by means of a compass placed on top of it? [Not definitely; only if the West end of the wire is positive will we get a deflection; a lack of deflection could also indicate a current with the East end positive]. What if we could place it on top and underneath? [Yes, the reversal of the compass from the top, say N-S, to underneath, S-N, would indicate a current].

Instead of fixing the wire and having the movable compass needle dance on an agate bearing, we could use a fixed cylindrical magnet and let a flexible braided wire, tied in a loose arc, spin around the magnet (El3). We experience the cable wrapping itself around the magnet, as if to form a coil around the iron core. We can also make a coil in another manner if we wind some wire into a helix; inside this open coil, we get much stronger fields than for a single wire, comparable to a powerful bar magnet. Making a coil of several turns objectively strengthens the effect; each successive turn is additive (El4a-c).

If we slide an iron bar part way into the coil, then activate the circuit, the coil will suck in the iron, as if by a ghostly hand. Simultaneously, the magnetic effect is stronger (El4d, e). The magnetism of the coil seizes and magnetizes previously unmagnetized iron. When it is drawn inside the coil, the iron comes under the influence of the coil—even extending its influence.

The various orientations a compass needle could have when placed around a coil are shown in El5. The pattern of filings is an image of the *possibilities* which exist in the space around a coil. If we placed a compass needle somewhere into this space, it would assume one or another of these specific orientations. Our iron filings are immediately magnetized by the coil; the neighboring N and S poles on each tiny iron filing attract each other, linking them up into a pattern which we can see. In drawing this enchanted image with its curved 'lines,' we have a pedagogical problem: Before us is the entirety of the imaged pattern; no single line is separate from it and stands alone. A person must have understood this *whole imaged pattern* or at least feel it as they draw the 'lines,' and some students can only do this with difficulty—naturally, we will work with them individually. This image shows only 2 straight lines, extending each way from the center axis of the coil. The curvature and thickness of the iron filing lines depends on the current strength.

Now, we come to the experiment which begins the whole of electromagnetics. The foregoing served more as a systematic grounding of teacher's knowledge; we will let the students simply be pulled into the experience and, at the outset, the reflections ought not be too challenging for them. We demonstrate the amazing forces a coil has, especially when the iron core is bent around the coil (El6). At the gap, its N and S poles lie very close together, which produces very strong forces. The yoke of this U-electromagnet can support a whole student, powered only by a small flashlight battery (El6d)!

This combination of copper and iron—the soft, friendly copper and the hard, dark gray iron as coil core—which through appropriate construction join to create quite large forces, is called an *electromagnet*. Each student can quickly build his own magnet with a wire and a large nail and experiment with it (El7). In technology, a somewhat larger cylindrical arrangement works well, for example with scrap metal on a dock, or where steel plates and other iron objects are to be moved about. At the press of a finger, a person can switch the force on and off from a distance, energizing the magnet so that the objects can be moved by crane and not by hand.

II. THE TELEGRAPH[2]

The fall of Troy was announced to Clytemnestra, the wife of Agamemnon, by means of signal fires (Aeschylus, *Tragedies*). In Schiller's *William Tell* drama, the uprising against the Guardians was similarly announced. The sky glow of burning villages and towns unintentionally became the proclamation of war or tumult. In such tidings, the dramatic character of the symbols still shares the quality of the content of the message. When storm flood-waters crest over the dike, the town clock can ring the storm alarm. **It is easy to signal things that have been previously arranged and are already anticipated by recipients; it is more difficult to explain unexpected news to distant listeners.**

Right up to our time, there has been a news runner for the Dali Lama in Tibet.[3] Herodotus describes the same system in ancient Persia.[4] For millennia, the only way to send messages involving a longer text was to inscribe the news, e.g., onto clay tablets with a writing instrument. The modern optical telegraph of Chappe (1793), a kind of flagged-windmill, was first used widely under Napoleon; in French Europe it was of decisive importance in announcing troop movements. A wooden signal flag positioned diagonally on a tower communicated letters of the alphabet in a visible code to the next station, perhaps 10 km further on. In just six minutes the letters of a message could be sent from Paris to Strasbourg—in clear weather. The word *telegraph* derives from this windmill apparatus: distant writing.

Soon afterward, someone thought of applying the newly-discovered phenomena of electricity, which could span great distances in an instant. In 1794 in Ghent, the Holunder telegraph was devised: a 24-wire connection (one for each letter of the alphabet) with a pair of Holunder 'signal balls' on the end of each linen thread. If an electric machine (static generator) made a spark on the sending end, the signal balls moved apart on the other end. However, it was just about impossible to keep the large electrostatic charge on these lines insulated over any meaningful distance.

In 1809 H. Sommering, an anatomy professor in Munich who had corresponded with Goethe about nerve impulses, discovered another impractical version: the electrode leaf telegraph. Again, a 24-wire connection was used, this time for low voltages. One end of the wires was dipped into water; at the other end, a voltaic column (discovered in the meantime) was connected to specific wires, which were electrified thereby. Another wire served for the return connection in the circuit. When the circuit through one of the 24 letter wires had an active current, then bubbles immediately began forming on the water-immersed receiving end, indicating the corresponding letter. Napoleon, who already had a flag-telegraph, laughed about the discovery and the vast stretches of 24-conductor cable between 'bubble-indicators' it could supposedly make possible. But, after his defeat, these 'windmills,' which had extended his power, were destroyed by the advancing English troops without a second thought.

After 1820 (Øersted), people everywhere began to experiment with electromagnetic telegraphs. In 1833 in Gottingen, Gauss and Weber built a needle telegraph in order to send signals within the city—entirely along the lines of our Øersted experiment El1. Repeated, rapid connection of the battery at the sending station produced a tracing by the deflections of the compass needle at the receiving station. Also in this decade, various needle- and indicator-telegraphs were developed, especially by Wheatstone in England. Much simpler, however, was the epoch-making armature telegraph of Samuel F.B. Morse (1844).

It would be useful to introduce an antique apparatus, replica, lecture-hall demonstration model or ship's 'sparker' if one is available (El8). Most printed Morse code tables have an error: the dot actually occurs as a short dash, and the dash or "dah" is equal to 3 such dots; a letter space also equals 3 dots, and a word space equals 7. The following illustration depicts a Morse telegraph station on land. Its technical construction was developed empirically (by trial and error in practice), discovering in the doing how the deeds and sufferings of electricity could be harnessed:

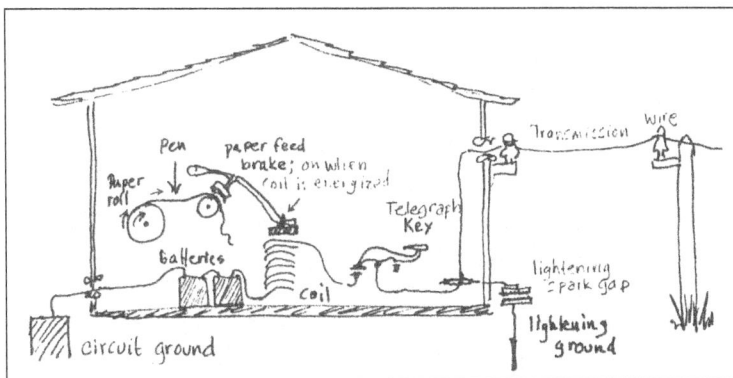

Figure 10. Morse telegraph station

As soon as we have built a regional telegraph station, we have a new problems: a) How many wires are required? b) How do we distinguish an unintentionally open circuit? c) How do we protect against lightning strikes? To answer a): People quickly found out that one pole of the battery could be connected to the ground; the broad, deep body of the earth can join in the circuit and replace one of the conductors, even up to great distances. It offers as little a resistance as a wire, as we can surmise based on what we learned in grade 7; a conductor has less resistance the thicker it is, and the earth is immeasurably thick It is astonishing that the earth can accommodate many such criss-crossing circuits without one disturbing another (see grade 7, on the isolation of each circuit). In this way, we utilize this isolation from the surroundings and adjacent circuits.

For problem b): We do not power the circuit with a working-current but rather with an interruption of the current, i.e., the armature is normally maintained under a current; it 'marks' only when the current is interrupted. The Morse key breaks the circuit for an appropriate interval. The first interruption sets the tracing paper roll in motion (takes off the brakes on the clockwork mechanism driving the paper roll). If the circuit remains broken for a long time—as for a line failure— then only a long dash would be written, the telegraph operator would know there's a problem, and repairmen would be called. By not writing, the apparatus gives evidence of a continuous driving current and the battery's charge.

For c): Protection from lightning can be achieved by a ground plate in close proximity (1 mm) to the circuit. It can have sharp ridges to allow a moderate voltage, albeit greater than the battery's, to jump across to ground.

Most Morse apparati function simultaneously as relays: The armature coil not only moves the writing apparatus, but a breaker contact lies in a separate circuit with its own battery, and though it, the received signal (break) is passed on to a further circuit or 'repeater.' (Why don't they just combine all these into one large circuit and have one master battery house? Generally, a series of small circuits is easier to build and repair than one vast circuit with much higher voltage—due to its increased resistance—and therefore thinner insulation.)

A dramatic chapter in the history of technology is the laying of a trans-Atlantic cable undertaken by Cyrus Field (1856–1863) with the first traveling steamer, the "Great Eastern" (1857, 19,000 Br. tons). Recent news and commercial connections have enthused people ever more deliriously since then. When the Atlantic cable first began functioning in 1856, a school holiday was declared in America. (A month later, the insulation was damaged and that cable became unusable.) As Charles Lindberg returned to Paris after his first Atlantic crossing, the crowds nearly tore his airplane to bits in their frenzy.

On the one hand, the connection between people is a deep necessity. On the other, the communications technology which has come to dominate today **fills people with a content which has not come from a direct experience of Nature or social intercourse**, but is more and more presented in the form of concepts. And these bare concepts, arising again and again out of dots and dashes, are half consciously shaped into suitable feelings by the journalist, since there is no such thing as emotion-free information. Also, the journalist cannot usually undergo the experience themselves, but collect them from other individuals. Our capacity for feeling, even for personal contact is pushed aside, and **passively-received mass feelings take their place**. Since such vast amounts of news rush in directly from many parts of the world, an individual seldom has a chance to focus on a subject for very long or even really 'live with it.' The feelings, which must be taken hold of to realize any human fulfillment, must now receive a daily recharge; constant stimulation makes it necessary. I know a frightful amount about distant places, but I cannot evaluate it and, therefore, my directly experienced human and natural surroundings deteriorate.

So, on the one hand, all of humanity can feel itself as a unity: Everyone has access to the same news, can see themselves threatened by the same random events. On the other hand, against distant, unstoppable realities, against the gigantic world we meet which we can no longer experience individually, nor feel our way into and thereby evaluate, there appears a basic mood—because now all this has occurred, a person is compelled to take it all up and deal with it. It breeds the sense that it is impossible, given where we are, to base one's life in any other way. Since people have become more solitary through the rootless news, i.e., more manipulable, previously people sought connections, conformity as a kind of action of self-expression. The large line? Already the modern younger generation, in many places seeks to take up new forms of life by pushing aside these technocratic forces. Will people create it, break free from the reign of terror of information outside and the craving for information inside, to create an experience of community and of Nature in ever wider circles?

Such views and similar ones should not be simply grafted onto the students. However, we could ask how life in our immediate surroundings would be different if a person—perhaps in the mountains—taught himself about new things by wandering throughout all the settlements.

III. THE ELECTRIC BELL

An electric bell affords the students a small exercise in training careful thought. In contrast to the Morse apparatus, the continuing up-and-down movement of the clicker against the armature coil must be created, not by a key, but self-produced; we should only have to press a button to start it (El9). It is often very instructive for the students to penetrate how voltage, current, magnetic field and mechanical movement work together here. This could be summarized in a main lesson book text, and a drawing something like this:

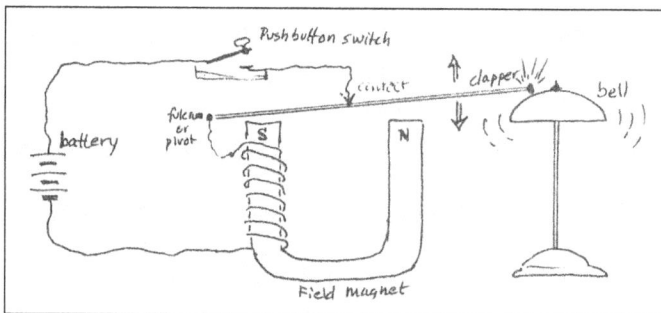

Figure 11. Doorbell ringer circuit

Another means of understanding the action is through the following series of drawings:

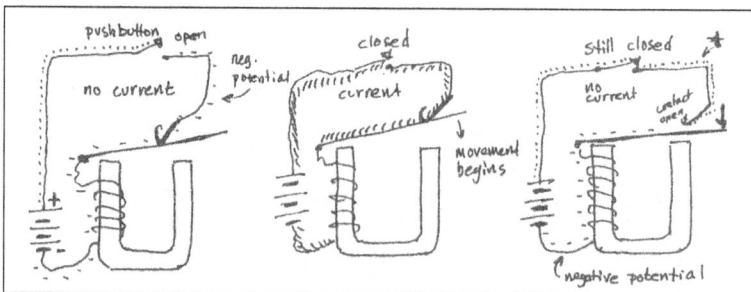

Figure 12. Operation of doorbell

We can mark the current path on the inked circuit—perhaps in green— by *current* we mean only the *availability* of a surrounding magnetic field (see El1); the *presence* of this segment of conductor is *a condition* for the complex of phenomena which we call current. If we remove this segment, then the magnetic field immediately ceases just as the anode in the battery stops being eaten away. The electric bell with the breaker contact was discovered in Frankfurt in 1839 and was found in the entrance to people's houses to ring them personally. The need for silk-spun wire (silk was the insulation) increased rapidly. Later came the bell on the front door. After about 10 years there was the AC current bell, which no longer had breaker points but depended on the alternating frequency of the current to determine the movement of the lever near the coil.

IV. MEASURING INSTRUMENT

In order to measure the strength of the current active in a circuit (the magnitude of associated phenomena), we can simply compare the strength of these phenomena: e.g., which wire gets hotter? We know that a thicker, better conducting wire barely gets warm, while a thinner one gets much warmer. Also, in order to establish the amount of heating for a particular current, we must utilize a unit quantity of wire; its thermal expansion can show the effect of the current directly. The sinking of the paper flag in El9 in grade 7 was such a current indicator. However, it shows a bit of a false reading, since when it would normally get quite hot, giving the indication of a strong current, the wire adds a tiny bit of resistance to the circuit, which reduces the measured current a tiny bit. Such an instrument also alters what we are trying to measure in a manner evaluable only with difficulty. Nevertheless, in early days, normalized circuits had such 'heater' or current-indicating wires (and even a long time later for specialized high frequency circuits).

We influence the circuit much less by measuring the phenomena of the magnetic field which surrounds the wire. We place a simple compass near the wire, and since the terrestrial magnetic field coincides with the wire's field, the needle moves to a diagonal position. The stronger the current, the more transverse the position. Such an instrument is called a *tangent galvanometer*. The measurement only works when the wire runs N-S and horizontally.

V. ELECTRIC MOTOR

We could spend a week of concentrated work on the practical aspects of motors. However, here we will only unfold the basic principles here, in order to clearly think them through.

What has to be modified on our rotating-coil instrument (galvanometer, El10) in order to enable it to perform continuous rotation? If we turn on a very strong current, the coil will clearly make a vigorous rotation, but only until it is aligned with the next magnetic poles—where it stops. The north and south poles 'stick' to each other after a 90° quarter turn. If the coil is disconnected at this moment, then due to its momentum it would turn on, past the magnet's poles, into the next quarter-turn. Now, the current has to be switched on again, but with reversed polarity. Where previously we had attraction, now we should have repulsion to push the coil along. By repulsion it performs another quarter-turn, and then it completes the fourth quarter-turn by attraction to the opposite magnet poles, where it would again tend to 'stick' to the poles. However, the same trick can help us: switch the polarity, and again every half-turn. Clearly, the connections to the coil must be designed differently: After a short turn, the insulated contact ring must be broken. Ordinarily, the supply current could be connected via two metal brushes which slide along two separate, insulated *conducting rings* or *commutator*. But who will make the rapid disconnection needed to reverse the polarity at just

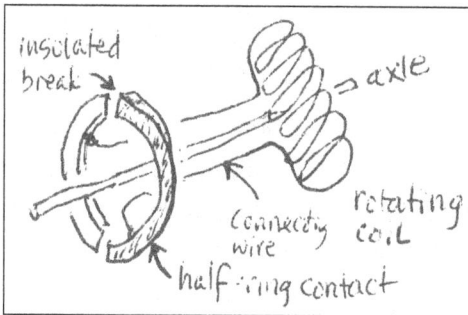

Figure 13. Single-wound rotor coil with two-commutator connections.

the right moment of rotation, just as it passes the magnet poles? We need an apparatus, turning at the same rate, which will break the connections; we apparently need the same motor we're trying to make as a polarity-reversing device! This dilemma can be resolved by placing the polarity-reversing part directly on the axle of the rotating coil, so that its rotation will provide the timed disconnection and reversed connection.

The rotor illustrated (the rotating part of the motor) has an axle in its center on which is mounted a plastic disk (the 'collector' or commutator). Along its circumference run two metal half rings, along which the current-carrying 'brushes' slide. From the contact rings, two wires connect the circuit to the rotor coil. This forms the entire reversing arrangement, including also the brushes. Today, these are no longer wires, but blocks of pressed carbon or graphite held against the commutator by wire springs.

How should the two brushes be arranged on the commutator? When one brush contacts one half-ring, then the other brush connects to the other half-ring or commutator contact, and a circuit is completed through the rotor coil, which becomes magnetized and turns toward the poles of the surrounding magnet (the field armature). How should the brushes be oriented relative to these poles?

Stage I: Connected by the commutator segments, the coil is under the influence of a current and acts as a rotating bar magnet; its poles are attracted by the nearby poles of a permanent horseshoe magnet (field), and it rotates in the indicated direction.

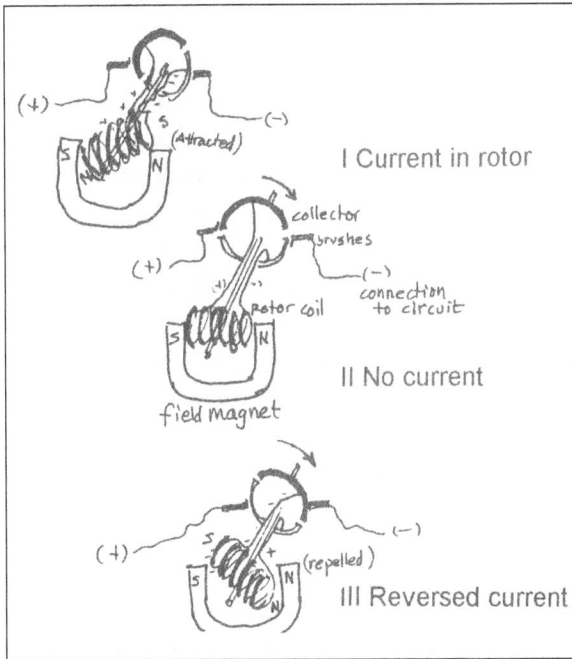

I Current in rotor

II No current

III Reversed current

Figure 14.
Relationship of commutator and rotor coil

Stage II: Here we have frozen the instant when the rotor poles just pass the field-magnet poles. This is when the reversal should occur; now the brushes run off the half-ring segments and onto the short insulated section of the commutator. These breaks should now lie parallel to the poles of the field magnet. The rotor coil has no current, is free of forces for an instant, and turns freely by momentum through this null-point.

Stage III: The brushes have made contact with the commutator segments again, re-energizing the coil, but with a reversed connection. The contacts are permanently connected to the current source, so the switched contacts reverses the current polarity through the coil. Where previously attraction acted between each coil end and the nearby magnet pole, now we have repulsion, causing the rotor to turn further.

These thoughts are verified in El11: There is actually a point where the rotor is current-free and therefore force-free and will not turn—the so-called null-point. We have to either push it to start the motor from this position or turn it on at another position in its rotation. Also, the sparking at the breaks between commutator segments (the 'disconnection sparks') demonstrate the constant disconnection and re-connection with reversed polarity.

To make the motor self-starting, we could cross-mount two rotor coils on a common axle, within one horseshoe magnet—a kind of dual motor. Now, if one coil is at its null-point, the other is at its maximum torque position. In practice, most motors use a multiple-pole construction in keeping with this idea; the various windings of the rotor coil are situated next to each other. Often, such multiple-pole motors have a 'barrel'-wound armature, where the conductors run parallel to the axle, with the polarity arranged so that each will repel or attract a surrounding pole appropriately; effectively, this spreads the coil out around the armature.

In order to make more and more powerful motors, we would have to use larger and more powerful field magnets (or "stator" in contrast to the 'rotor'), which would make the motors too large and heavy. This can be avoided if instead we use a multiple-pole electromagnet stator coil around the rotor coil. All modem motors are built this way, with a stator as well as a rotor winding. A miniature permanent magnet motor could easily be made by each student at home (El12). In addition, we could open up a hand drill or a blender motor and point out the graphite brushes, where the greatest wear occurs in the motor, requiring this part to be replaced on occasion.

VI. THE DYNAMO (GENERATOR)

Lastly, we can introduce the generator principle: voltage production (current production) through mechanical rotation—as with the Faraday dynamo. Any electro-motor with an iron core rotor (not the so-called shunt-wound or bird-cage motor) creates a voltage potential in the windings if it is forcibly rotated (El13). The associated details of generating current by induction are deferred for high school physics (grade 9).

VII. THE ELECTRIFIED WORLD

Instead of a continued "*scientization of all domains of life*," for the first time people today are coming to the awareness, as we all have in this decade, that the industrialized and technologized world still has bypassed or surpassed the essential issues of raw materials, pollution, energy—to the point where it can no longer remain like this. We are overly technologized; unnecessary waste has become everyday business. Almost all young people acutely sense this today, not because they know all the facts and figures intellectually, but because in their search for a life based on human values, they experience this excess technology. Electricity is the actual bearer of this growing technology, and with particular ease it has created conditions hostile to humans in three main directions, which we can discuss as part of our reflections on the telegraph:

First, in social realms, we see in radio and television and the supporting electronics something which separates human beings, turning them into individual units. **A totally different connection arises between people if, through personal conversation, they discover those things which are essential and enriching, directly from one another—not through mass media.** In the intellectual sense, the information may be the same; but for social communion, for humanizing of the feelings, electrical communication makes us poorer. Much is the pity.[5]

Secondly, in terms of the world of work, most of the time in the urban environment there are no longer many, if any, **fulfilling tasks** for us to do with our hands. The self-sacrificing concentration of bodily work, connecting the human spirit to the world by means of an elemental objectivity of the will, no longer has a rightful place. Only a diminishing number of professions can still "work" in this way. And, sports, hobbies, and vacation are no substitute.

Thirdly, when we are cut off from real social intercourse (feelings) by the first complex, and from meaningful work (our will) by the second, we are made **foreigners to thinking** by a third. The style of thinking of an electrified culture is the formal, additive of yes/no, switching on/off or in/out. For this reason, the computer is well-suited to our time, since we ourselves certainly use this kind of formal-logical thinking and deciding in all domains of life, i.e., formally combining concepts in a yes-no logic, without penetrating or experiencing the qualitative of their contents, and thereby creating something new.

In grade 8 we can scarcely lead students to a fully independent evaluation of the new technology and commercial developments, as I have tried to do here. My comments are intended only as a stimulus for the teacher, to look critically on the 'achievements' of electricity. And, by means of concrete, individual questions, using whatever practical points the students may place, try to foster the mood of a freely-willed limiting of technology, initially on a personal basis. For that is the task of humanity today—*to allow such issues to come to life in conversation between people once in a while,* not out of all sorts of theoretical foundations of Futurology and the Club of Rome, not out of the large overview, but rather out of **the careful sense in each of us for the rightful place of the human being**. Then a person might feel differently than Faust's assistant Wagner, with his naïve love of the head-knowledge, that Faust had found so hollow: ".. and how far glorious progress has been brought." (*Faust*, Part I, Night, Ln 573)

Experiments in Electricity

EL 1 COMPASS DEFLECTION (ØERSTED'S EXPERIMENT)

An insulated hookup cable with banana plugs (or alligator clips) is stretched out in a nearly N-S direction, suspended about 50 cm above the table top:

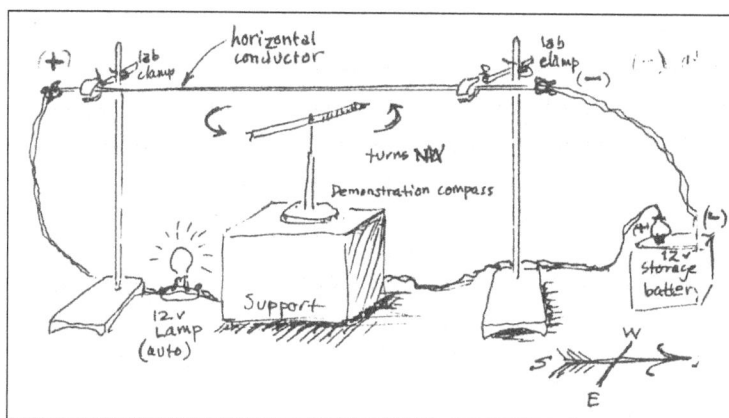

Figure 15. Compass deflection near a
current-carrying conductor

a. Switch on the current in the wire, perhaps with a lamp in the circuit to show the presence of electricity. A 'demonstration compass,' supported on a board, when brought close by the wire (N-S), will show a deflection, diagonal to the wire (approx. NE-SW).

b. Increase the current, to the point at which the cable begins to heat (use a storage battery) and again position the compass needle near the wire, both above and below: A stronger deflection is seen. Now the needle approaches a deflection which alternates almost 180° from above to below, although it never fully attains it. This effect is diminished for a larger compass.

c. Instead of the cable, we could insert a metal rod (lab stand rod or, better yet, non-magnetic aluminum or copper tubing) into the circuit. Experiment further how when the current is switched off, the compass again points north. (Note for teachers: Reversal of the polarity does the same as placing the needle above and below).

EL 2 CIRCLE OF COMPASS NEEDLES

Glue a tube of aluminum, tin, or rolled sheet steel, about a pencil-lead diameter, into a hole in a 20 x 20 cm square of thin plywood or stiff cardboard; paint the sheet white on the upper side. Now place about 6–12 compasses around the tube/rod.

A battery is connected, sized so that the rod/tube does not heat too much. The compasses shouldn't be too close to each other, or they will simply be attracted

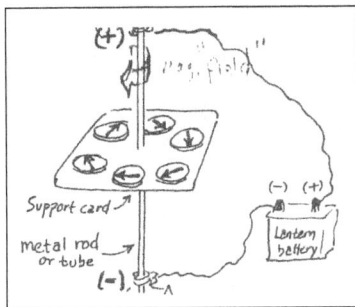

Figure 16.
Compass deflection pattern
around a conductor

to each other, instead of all pointing North. Place some further from the conductor, near the edge of the small shelf. When current is switched on, the compasses are all deflected, particularly those located just north and south of the conductor. They all tend to form a circle around the conductor, less so for the ones further away. Reverse the storage battery, and the compasses form a circle, going the opposite way. (Each time the circuit is switched on in a new way, I have the students all come up and look).

EL 3 SELF-WINDING WIRE

A bolt of #59 tool steel, 4 x 20 cm, has a 5 mm slit on each end with a slanting pin in each, around which we can tie a wire. The bolt has been magnetized by hammering on the bolt, while in a coil of about 250 turns with 10–20 amperes DC current; the steel bolt is then wrapped in tape to insulate it, and thick, insulated copper wires attached to the end pins.

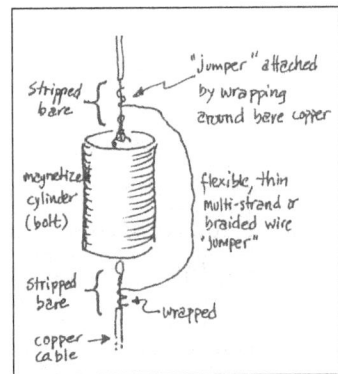

Just beyond the ends of the bolt, the thick wire is stripped of insulation; we wrap the ends of thin, flexible wire (multi-strand, 0.1 mm² diameter) around the bare sections of the copper cables several times, to prevent it from sliding down over the bit of insulation left on the large wire.

Briefly apply a strong current to the copper wires, and thus to the thin, braided or stranded wire 'jumper.' The thin jumper will heat up, and also attempt to wind itself around the cylindrical magnet of the bolt. The experiment requires luck [not to overload the thin jumper, but to have the magnet and current strong enough to produce sufficient force to get the jumper to move].[6]

EL 4 COIL AS A MAGNET

a. A stand-mounted (demonstration) compass is allowed to settle into a N-S position; it is held up on a small board, mounted securely with a lab stand and clamps. Similar to El1, we show how a cable, powered by a 12 v storage battery and held above or below the compass will produce a deflection.

b. Make a coil of about 6 turns of multi-strand thick wire and large enough diameter to enclose the compass needle (wound around e.g., a cylindrical cardboard oatmeal container). Mount it so it can be held around the mounted compass.

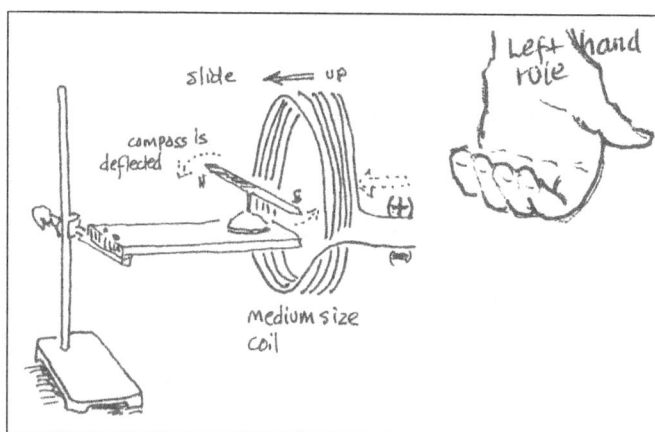

Figure 17. Strong deflection by a coil

c. Place the compass on the table; show deflection of both its north- and south-seeking ends, first using a bar magnet held a few feet away, and then using a coil of about 250 turns (use first one end of the coil and then the other). If the current in the coil is increased just short of the level where the coil gets hot, a magnetic field will be created, just as if a bar magnet ["invisible magnet"] were situated in its open core, along the axis of the coil. The coil should be as large a diameter as possible, with a 4 cm diameter opening.

d. Slip a matching iron bar into the open core of this large coil while it is switched off, but not all the way in (not symmetrically into the coil). Switch on a strong current, and the iron core is suddenly pulled into the coil as if by an invisible hand.

e. Repeat experiment El4c with this iron-core coil. We get a much stronger deflection of the compass.

EL 5 FIELD AROUND A COIL

Using the large coil from El4, mount a paper cut-out around its equator. Sprinkle fine iron filings over the paper and switch on the current (using a 12 v storage battery or power supply). The filings align themselves into a pattern of 'lines' and strips, concentrated around the coil openings. Lightly tap the paper to aid them in taking up their pattern. Quite a few filings will be sucked into the coil openings (like the iron core in El4d).

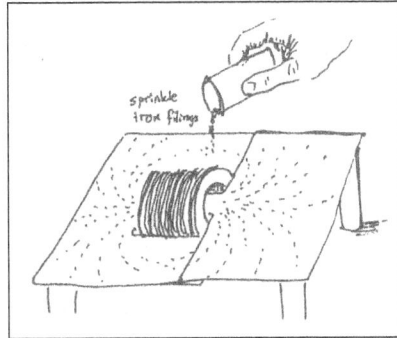

Figure 18. Field made visible

EL 6 ELECTROMAGNET

Some effective force experiments are the following:

a. A coil is (briefly!) energized with a current just strong enough to cause warming, and held up in the air. The iron core will be suspended, floating inside the coil. It bounces with a movement of the coil; when the current is switched off, it falls end-down with a crash, onto a metal plate placed below.

b. Needles, spoons, etc., stay 'stuck' to this electromagnet when they are thrown against it. Alternatively, they are 'sucked up' from the table by the coil, and fall down when it is shut off.

c. Slide the coil onto a U-shaped core, energize the coil, and try to slowly bring a steel bar (a 'yoke') against the ends of the u-magnet. It will suddenly 'jump' the last few millimeters and crack against the magnet poles. Have a student try to pull the yoke off while the current is on: a futile effort!

d. Our electromagnet is a u-armature with two coils, one on each leg. Suspend it from the ceiling and power it with a 1.5 v battery. The iron yoke has a hook for suspending a student on a rope seat (see mechanics, grade 7; student wears a motorcycle helmet). This will support most students, although when the battery is used up or the student wiggles, he will abruptly drop. This silent falling (or disconnection of the battery) is so insidious that we could test whether the yoke is attached by delicately bouncing it.

EL 7 STUDENT ELECTROMAGNETS

Wind 60–70 m of insulated bell wire or hookup wire on a wooden speed-winding armature with a dowel axle:

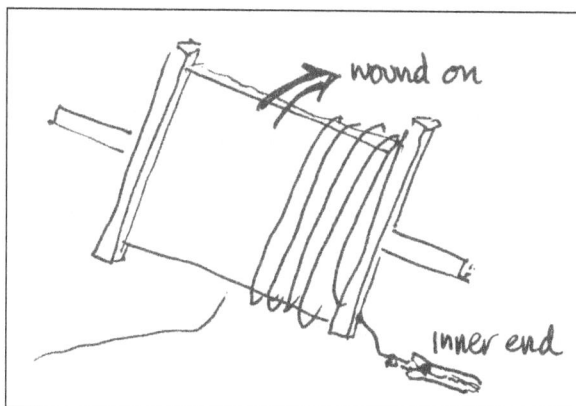

Figure 19: winding armature for magnet coil

As shown, the inner end of the wire (the other windings are not shown) should lead to an exposed connector. Now, we go up and down the rows of desks, around the class, unrolling the shiny copper loops from the armature, until each student has unrolled at least a meter. Plug the ends into the DC power supply (or battery) and turn up the current. The students feel the heating, indicating a current. Now, they each wind about 20 turns around a pencil; investigate (with and without a pencil) how the coil attracts small brads. The magnetism is too weak. A pocket compass placed nearby is also hardly affected.

If they now slide a large framing nail ('spike') inside their coil, then brads and even iron nails will be easily attracted, often in long chains, and even magnetized. The large nail remains magnetic afterward; it can be demagnetized by hammering, heating red-hot, or inserting it within a coil energized with alternating current for a few minutes. When the long wire is disconnected, all the nails attracted to coils throughout the class will simultaneously fall away.

EL 8 TELEGRAPH

We set up an experiment using an antique Morse key (or else, rigged up with a single-pole, single-throw, normally closed pushbutton):

a. If we bypass the key (switch) with a wire jumper, and then manually disconnect the jumper, the clicker snaps away from the coil, and the paper is free to unroll underneath the pencil (see Figure 10), drawing a long line.

b. Now, reconnect the sending key into the circuit, depress the key (closing the circuit,) and show how a finger's pressure is sufficient to open the key-contact, so that the paper unrolls as above.

c. Connect a relay into the circuit so it will also be energized by the Morse key and its clicker coil circuit. See how this separate, secondary circuit (the contacts of the relay) is switched on when the Morse key is pressed, as shown using a small pocket lamp, energized by a flashlight battery.

d. The key could be placed in the rear of the classroom, connected with 2-conductor wire, and messages sent to the front, deciphered at the clicker coil there.

EL 9 PRINCIPLE OF THE ELECTRIC BELL

We set up a Chladni plate and a bow-saw blade mounted above it, with a few heavy crocodile clamps to adjust how hard it presses against the plate. Bring the bow saw to ringing, like the Chladni plate, with our rosined bow. It is important that the saw blade not only vibrates, but really oscillates widely (whips)—this may require a bit of experimentation. The breaker contact (loose contact draped from the tip of the saw) must be elastically mounted to touch the plate, so that it can follow the motion of the saw just before it breaks. At this spot, the plate surface should be rubbed clean. A banana plug on a rubber band is the breaker contact; it is clamped from the rear by a lab stand clamp, so that the metal rod of the banana plug can move back and forth with some springiness. (Alternatively, use a hair-thin steel wire.) First, show the oscillating without the plate. Then, connect the circuit and show how the oscillating makes an oscillating interruption in the circuit, making a bell 'ring.'

[Alternatively, use a regular household door-bell. First, show how it 'clicks' if connected only by the 'ground' (on its frame) and one of the 'coil' contacts; i.e., the bell clapper just clicks down or away as the circuit is made and broken. Then, wire it in at the usual terminals, showing how these will make a circuit through a tiny set of contacts attached to the clapper, which open and close as the clapper clicks down or away. When energized, these move rapidly with the clapper, breaking or making the circuit, causing the bell clapper to buzz and the bell to ring. – Transl.]

EL 10 ELECTRIC METER (ROTATING COIL)

Make an air-core coil using 6 turns of thin, stiff wire, having a good deal of space between each turn; suspend it on flexible metal bands (leads).[7] The coil is capable of turning between the poles of a large horseshoe magnet; however, the tension of the suspending bands will turn it back. This mechanical tension will determine the sensitivity to an applied current. A long pointer (straw) and a large, clear scale complete this current-measuring instrument.

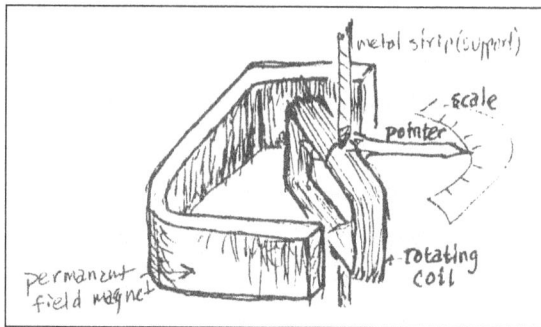

Figure 20. Measuring instrument

The deflection is proportional to the brightness of a bulb in the circuit (i.e., the magnitude of the current). Reversing the polarity produces a reversed motion.

[Note: the tiny coil inside a galvanometer (or almost any meter) can also be used to show this, although less clearly since it is so small. – Transl.]

EL 11 ELECTRIC MOTOR

Use a demonstration motor, initially having a light bulb and battery wired into the rotor coil circuit; we show how the brushes and commutator function, i.e., at which position in its rotation the current acts in the rotor and for which the current circuit is broken (at the joints in the commutator). Using a compass, determine the polarity of the rotor coil at these energized positions, e.g., when its coils point up/down; or, insert the rotor between the poles of a strong permanent magnet and note the polarity of the coil within this surrounding magnetic (stator) field, i.e., which end of the rotor coil is north-seeking and which end is south-seeking. Now, position the rotor poles (as determined by the compass) transverse to the stator magnet poles. From our knowledge of the polarity of the two pairs of poles, we should be able to predict which way the rotor will turn. Switch on the current, but keep it very weak at first (using a variable resistor in the circuit), so the rotor barely turns. Then, increase the current, allowing the rotor to begin turning more and more rapidly. Observe the sparks formed by the brushes as they turn away and disconnect from each section of the commutator.

Now, show how the motor won't start if positioned at the null-point (N-seeking S rotor pole at the N field pole, etc.). Contrast this to a multi-pole rotor which always has a pair of rotor coils positioned ready to move. Finally, replace the permanent field magnets with a U-shaped electromagnet (if your kit allows). Since the field (stator) poles of this electromagnet are curved, they don't only act on a point, but enclose a whole segment of the rotor's rotation. This helps prevent the rotor from stopping or going backward, since the rotor's coils already experience an attraction, even while they are still angled quite a bit away from the center point of the stator pole. This is how most modern motors are constructed, although the details of how the stator coils are wired into the circuit may vary.

EL 12 HOME-BUILT MOTORS

(Description of various complete motor kits, available in Germany. See the PSSC section in Sargent-Welch, Fisher Scientific, or Frey Scientific supply catalogs for examples of these student motor kits). However, in my experience with such kits, there is always the "Monday morning glitch": One student will have a part that did not quite fit and we may need to return it to the company for a replacement.

EL 13 DYNAMO/GENERATOR

For this, we can best utilize the demonstration motor with barrel-wound armature. It can be turned rapidly, using the hand crank we used for Grade 7 acoustics, Ac7 (using perhaps a larger rubber-edged driving wheel against the axle of the motor). Via two banana plug connecting cables, we power a smaller secondary motor—one motor drives another! Or, alternatively, using only one motor, we can light a light bulb.

Grade 8 Hydraulics/Hydrostatics

INTRODUCTION

Rudolf Steiner's indications for this study are as follows: "In grade 8, again we extend by recapitulating those things which were cultivated in the 6th grade, and move on to hydraulics and the *forces which work in water*. Also, take up everything which has to do with the concept of *buoyancy*—such as pressure exerted in all directions and everything connected with Archimedes' principle, which also belongs in hydraulics." Hydrostatics also comes into consideration: forces in motionless water. Clearly, with [still] water [hydrostatics] we shall focus more on the dead aspects. Nevertheless, a deep inner interest of the 'sense for reality' will be brought to bear only when we treat water in accordance with a sense for its role in the natural environment—not only in keeping with its aspect as a solid, but rather fundamentally in accord with its fluid nature.

This latter is foreign to classroom-physics: We will have to rethink it. The important thing, for example, is that *in water, water has no weight*! I have to change it to a solid condition—enclose it inside a container—if I want to weigh it. When I pour it into the ocean, once again it loses its weight. This is not to deny that, in addition, we can think of water in the ocean using concepts of weight. Nevertheless, in contrast to chunks of wood, stone, and the like, water phenomena bring something else to the forefront: namely, *pressure and fluid flow*. Thus, bodies containing only a small amount of water will also become somewhat lighter in water, even if they don't lose their entire weight as a whole, as their watery portions do.

We won't focus initially on the phrase learned in school about buoyancy related to displaced weight or water, etc., but will first seek a practical approach with what is closest to hand. We will not attempt to put water into a solid state (using a container) in order to weigh it, but rather first *immerse solids into water*, measuring, for example, the pressure. (Note that air also belongs to this group of "solids," in that we can exchange it for a solid and pressure is measurable). If I put a boat in water, before we measure the water displacement, we must consider whether the boat hull is thick enough, or whether the pressure deep below at the keel will make a leak in the hull plates. Everyone knows how dangerous a leak in the hull is; water will squirt in like a powerful natural spring or geyser. Above, up on the wall of the ship's hull but still deep under the waterline, water will still try to run in through every crack, but will not squirt as far and can be more easily plugged.

In this way, we see that the phenomenon in water is pressure, not weight, and from these pressure-phenomena we will take our start. In light of this, we could review Rudolf Steiner's indications, above.

I. PRESSURE PHENOMENA

The term *water pressure*, or more generally, *fluid pressure*, denotes a concept which is often not clearly understood. Before we investigate how pressure increases with depth, we should learn to understand pressure itself—as a varying characteristic of enclosed fluids. Why can I squirt farther with a thin plunger than with a large-diameter one—as seen in demonstration Hy1? The distance we can squirt is clearly an "ex-pression" of the inner situation in the fluid, of its out-flow force. The greater the pressure, the more the water seeks an outlet from the enclosure, and the greater the velocity of outflow. The pressure in the plunger cylinder pushes mostly on the walls, and—as the multi-direction squirter shows—presses equally on the side walls as well as on the bottom wall. Here, pressure is non-directional: it is *a condition of tension on all sides which the place "wills" to create equally in all directions*. This "will" naturally does not arise from the water, but out of what creates the pressure; this non-directional force is simply passed along [through the water].

First off, many students may mistake this squirting for a movement-phenomenon, thinking the movement of the plunger is simply passed along into the water coming out the nozzle (exit tube); in an obvious way, its movement is continued straight ahead. However, this is contradicted by our observations with the multi-direction squirter (angle squirter Hy1). It is necessary to understand all these experimental observations *not* as a continuation of the motion of water driven in a particular direction, but as **pressure phenomena**. The water pressure of the piston, or any strongly compressing, pressing-together, squeezing-in motion creates a condition of tension, called *pressure*. Pressure is exerted outward, equally in all directions. Regardless of the side where the pressure arises, it will be exerted completely equally to all sides. Wherever an opening exists in the containing walls, the water squirts out there—equally strongly in any direction. The direction of the initial force or push can be diverted to a transverse or even the reverse direction;

although certainly only for pressures which can be withstood by the containing wall. This reminds us of the solid roller, studied in mechanics last year.

For every jet, the applied force serves only to place the water under pressure [not to push it in any particular direction]. At the bottom of a large flask, less pressure will exist than in a smaller flask. The containing force of the flask corresponds with the water pressure by means of the enclosing surface; this surface is where the enclosing force passes over into water pressure. For a large flask, the pressure of the piston moves equally along the neck (with a constant diameter), but spreads out in the wider round part, and is divided up over the surface of the flask where it presses into the water. Since the given force is apportioned over a large flask, it cannot press as hard; i.e., the water is not placed under as great a pressure. For the tiny squirter, things are quite different: The force is concentrated on a smaller piston surface, so the smaller surface has to contain the entire pushing force; thereby the orifice receives more pressure. Mathematically stated: In the small flask, the ratio of force to surface-area is much greater than for the larger. This condition of tension (*pressure*) prevails equally throughout the entirety of the fluid and also at the nozzle, where the water squirts out at a velocity depending on the pressure. The pressure acts the same on the glass wall; it must contain the pressure. The tiny tube of the jet hole(s) is like a 'pull'; naturally, when there is no containing pressure, the velocity of the water streaming out of the nozzle corresponds to the pressure in the flask.

If we put a person *under pressure* in order to get them to do something, it is similar to the situation with the water in the squirter: We push from one side to get them to do something toward the other side.

In a volume of a fluid, the force spreads out equally to all sides—also through a tube. The force applied to one end can be found again at the other. Thus, we can use two study plungers with equal areas as a hydraulic balance (Hy2). Two unequal plungers direct us to the possibility of the hydraulic press, the hydraulic lift, etc. The smaller piston can create great pressure with little force; the larger piston, by virtue of its larger surface area, takes hold of a much greater pressure in the fluid (water or oil, Hy3). We could think of the larger piston surface as divided among two smaller pistons. The forces are proportional to the surface areas, and the distances traveled by each piston are naturally inversely proportional to their piston face areas. If we move from an equilibrium of forces over into movement, then the forces on the larger piston are a hundred-fold larger, but the travel is only a hundredth—like a lever. We are familiar with such a hydraulic translation of force from its everyday application: the hydraulic brake in a car. The small force on the brake pedal is translated into a larger force (and small movement) on the brake shoes at each wheel.

If a person wants to express the pressure in a formula, then we have to say: Pressure equals force divided by surface area (e.g., at the small piston); force equals pressure times surface area (e.g., the large piston). Previously, this was measured in straightforward units of kg/cm^2; where $1\ kg/cm^2 = 1$ atmosphere, usually approximating atmospheric pressure. Today, the metric system of measurement (the SI, System Internationale) prescribes the base unit of pressure as the *bar* with units of: N/m^2 (Newtons/square meter). One bar = $105\ N/m^2$ = $1.019716\ kg/cm^2$ (at sea level and 45° latitude).[1] One must feel out whether such Formulas should be mentioned to the class. We don't bring about new concepts by such formulas, therefore they could intimidate those with a feeling-orientation and empower those of a more intellectual one.

In the preceding, we have learned about forces which act in water. Such forces come to manifestation *wherever the water ends and the container wall begins* (the wall confining the water on all sides). In order to use these forces, a part of the wall must be movable, in relation to the surrounding wall. A piston in a cylinder is very convenient and works well. The piston should be quite thick so that no water leaks out and thus reduces the usable force. For example, an excavator (backhoe) moves its hydraulic shovel by means of oil cylinders, in which the piston is sealed by rubber seals to prevent leaks.

Within a fluid, pressure prevails—except we cannot see it. We have to imagine something quite undirected, unformed there. A very *dim feeling of being oppressed and under tension* is the only analogy by which we can think of it objectively. Clear concepts of force-direction, potential for movement, etc., first appear only when we reach a boundary, i.e., when we pass from a fluid into a solid condition.

II. DEPTH PHENOMENA

Hydraulic motion, if it is to be applied, occurs more as a phenomena in technical apparatus. In that case, the fluid moves over a bare surface, e.g., a piston face, driven by high pressure. In Nature, fluid *pressure appears only in the depths* of bodies of water; however, it does not act on the walls (the slanting bottom), but on individual bodies in it, e.g., a floating person. At a depth of one meter, we notice the pressure mostly in our chest cavity and in our head. Why especially in the chest? Here we have brought a little bit of the upper world with us down into the depths: a small volume of inhaled air. This will be compressed to a smaller volume (gases are quite compressible; fluids are a thousand times less compressible, though still compress a tiny bit!). Most body parts, e.g., arms and legs, won't experience the pressure, since they are made of moist tissue and solid bone. It is the same for the fishes, especially those without a swimming-bladder, e.g., the sharks. They don't experience any pressure since they are inwardly mostly fluid and their tissues are not at all rigid. *Fluid things do not actually sustain any pressure in water.* The influence of pressure is felt first when something with a solid boundary goes down into the depths, e.g., an air bubble or an air-filled lung or a submarine with steel walls. In this way, we have measured that pressure increases in water in almost exact proportion to the depth—up to 1000 atmospheres (approx. = 1000 bar) in 10,000 meters of water.

With a pressure-flask, we can measure the pressure at various depths in an aquarium (Hy4). The pressure laterally and upward is the same! It doesn't depend on the orientation relative to the wall nor on the form and size of the water container, but the magnitude of pressure depends only on the depth at which we measure it, e.g., at half the depth, we get half the pressure.

For the "Pascal's vases," the apparatus to measure the pressure becomes a kind of water tank; this is a bit of a contrivance. The measurement of hydrostatic force in Pascal's vases is practically a measure of weight or downward force of the water poured in, and by itself, this experiment could be misunderstood as dealing with weight/force instead of pressure. It would be more clear if a person could attach the apparatus to the bottom of the aquarium tank. This is clarified by the liquid pressure measuring apparatus (or pressure U-tube manometer) in Hy4. Again we see that pressure doesn't depend on the quantity (volume) of water, but only on the height of the water column and the area of the opening. This is also shown by the communicating tubes (Hy6) apparatus.

Alternatively, if we connect two funnels with a length of hose, we can make a flowing water-circuit and use it to determine if, e.g., the foundation of a house is truly level. Other examples are how fluids pour out of the spout of a tipped coffee urn or in the siphon built into a washbasin. All these examples are based on the two phenomena: (1) that water pressure acts in all directions and (2) that the magnitude of pressure is determined only by the height of the column.

In water, the pressure increases one atmosphere (about 1 bar) every 10 meters. At the surface we have an atmospheric pressure of 1 bar; further down, the pressure is augmented by the depth of the water. This can be illustrated nicely by taking a cannister (Mariotte's cylinder, Hy7) with holes in the side and immersing it empty in the aquarium; the water coming in through the holes will make correspondingly stronger or weaker streams depending on the depth, just as when the container is filled and the water streams out.

This increase in pressure with depth in water can be restated: The deeper under the water I bring a quantity of air (like the measuring tube in Hy4 or a submarine), or the higher I lift water in a container above the earth's surface, the more it tends to exclude this displaced space (compress the submarine) and the more it tends to flow down and rejoin itself (pressure increase on the container bottom as it is pulled upward out of the water). In this pressing action lies the tendency to fill out the depths of the earth and round it off.

III. FLOTATION/BUOYANCY

If we immerse an empty watering can in a water tank, then we sense an upward pressure or buoyancy with which the can is pressed upward. We imagine a body with vertical side walls and a horizontal bottom surface set down into water. On its lower surface, there will be an upward pressure, corresponding to the depth, and an opposing force of its own weight downward.

How can things float? Imagine first a wood block laid in water, which we know won't sink.

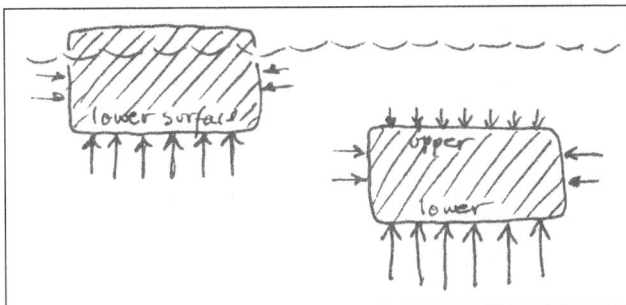

Figure 21. Bouyancy as an expression of depth-pressure

In the air, it must have a supportive upward force in order not to fall. Simply: I have to support it with an upward force if I wish to hold it in my hand. This force now has to be created by water as a pressure on the bottom surface of the block. The requisite supportive force which we are familiar with, which balances the weight [downward force] is provided by the water, or else the block would sink. The pressure on the lower surface must also produce a force which supports the block.

As the pressure increases with depth, when the block sinks deeper, it would be pressed up again, since its lower surface had come into a region of greater pressure. Conversely, if we lift up a body swimming in water (floating in equilibrium), its bottom comes into a region of too low a pressure, the lifting force now can no longer support it, and it will sink back down. Thus, if a body can float, there is a definite, stable situation in which it will float, with a certain water "draft."

When does a body sink? When it sinks deeper and deeper into the water and still doesn't receive a lifting pressure/force from the water, because it is too heavy. Going still deeper doesn't help: The pressure certainly increases on the lower surface, but once submerged, a downward pressure nevertheless now acts on the upper surface. This latter pressure counteracts any increase in pressure experienced on the lower surface.

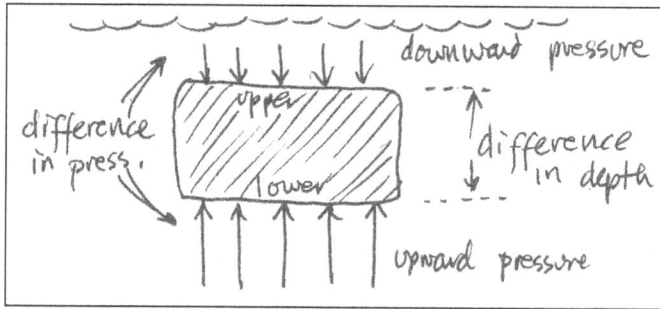

Figure 22. Pressure difference top and bottom

The difference between the lifting pressure and the downward-directed pressure is the same for all depths (since they differ by the thickness of the block, which remains constant).

If we consider for a moment the instant of submergence, above the body there is as yet no water. On its lower surface, the water only lifts upward. In order for the body not to sink deeper than this point of being just awash, the water must exert on the whole surface of the body exactly the same lifting force which we exert if we hold the body in dry air.[2] How great, then, is the lifting force in water at, say, 10 cm depth? This has to be exactly measured (see Hy4), and we find: 10 grams per

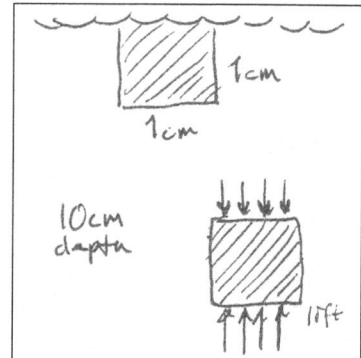

Figure 23: Effect of depth

cm². At 10 cm depth, water supports an immersed body with 10 grams/cm² on its lower surface. If the body requires a larger supportive force (it is heavier), then it will sink. If it requires less than this, then it will be lifted up until it emerges from the water. In order to arrive at the volume-weight, think of a cube of 1 cm immersed 1 cm deep.

The lifting force on the lower surface is 1 gram since at 1 cm depth, the pressure is 1 g /1 cm². If we enlarge the body to 2 cm³, then there are two ways it could be submerged. (That only one of these is a stable configuration plays no role here.) For, (a) due to the doubled submergence, a pressure of 2 grams acts; for (b), only a pressure of 1 gram, but on a doubled surface area of 2 cm². We already suspect the important experimental fact (Hy8): The lifting force (buoyancy) depends only on the submerged volume of a body, and the way it floats depends on its form. If its weight is more than 1 gram for a volume of 1 cm³, then the depth-dependent buoyancy no longer will suffice to buoy it up.

Materials whose weight per volume (also called: specific gravity) is greater than 1 g/cm³ will sink; those with less, will float partly above the water's surface. The specific gravity of water must be 1 g/cm³, since it is supported by water.

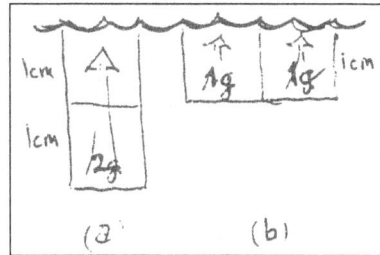

Figure 24.
Effect of on larger block

If we take a water column of 1 cm² area, e.g., in a 10 cm high cylinder, the pressure on the bottom = 10 g. The lateral pressure on the rigid walls of the cylinder is resisted. Also, the

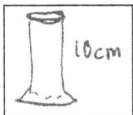

cylinder weighs exactly 10 g more with the water than without. Seen this way, water has a specific weight of 1 g/cm³. We can also say: Only those bodies with a density less than water can float.

The lifting force (buoyancy or decrease in weight in water) is always just as great as the weight of the displaced water (Hy9). This is Archimedes' principle. As the submerged body becomes "lighter," buoyant, the whole water system must get correspondingly heavier. (The load is transferred from our hand to the tank.) The consideration of pressure leads quickly to this: As the water surface rises as a body is immersed, the container bottom lies under a correspondingly greater depth of water and therefore receives proportionally greater pressure from the water it contains.

If basing of buoyancy on pressure, above, doesn't appear important, students could begin with this experiment (Hy9). For swimmers, this is naturally significant: A ship sinks to the point where its displaced water equals its own weight (Hy10). The submerged walls of the ship have to hold out the water pressure which arises from depth. At a 5 meter (16.4 ft) depth, the deepest parts of the ship hull have to withstand about 0.5 kg/cm^2, i.e., 5 tons/m^2!

A well-known example of hydraulic pressure technology is the water supply system for a small town or village. Taking a look at each component of this system, we can readily explain how the water comes out of the faucet.

Questions for reflection to deepen what we have covered so far:

1. A canal leads over a street via a canal-bridge: A driver in a car sees a boat traveling along the canal above. Should he wait until the boat has passed overhead? Does the canal-bridge bear a greater weight than at other times? [No. The water pressure is only a function of the depth, independent of a body floating on it; or, from the point of view of weight: the boat (metal hull, deck, etc.) displacing water and floating, exerts just as much pressure on the bottom as the water that would be there if the boat wasn't.]

2. Will the canal-bridge be more heavily loaded if a boat sinks in it, i.e., lies on its bottom? [Yes. The weight of the boat parts must now be added to the weight of water in the canal. Note: the submerged hull deck boards, etc., are slightly lighter by the amount of water pressure on the parts due to displacement, about 1/6 their weight (the inverse of the density of steel). For a heavy boat of steel, the additional weight of the sunken metal parts (5/6 the dry weight) almost equals the water previously displaced by the boat when floating.]

3. In the channel of a locks, a ship lowers a rowboat into the water. Does the water surface rise? [No. The water, which is displaced by and bears the weight of the ship and rowboat separately, previously carried the weight of the ship and the weight of the rowboat as one.]

4. Some of a load of stone slips from a heavily laden ship into the water of a lock channel. Does the water level rise in the locks? [No, it falls. The pressure of the stone (indirectly) upon the water is gone: it now rests on the bottom and is supported by it, rather than being carried by the buoyancy due to the displacement of the ship's hull. To equal the previous situation, a greater quantity of water would

have to be displaced than just the volume of the stone; but this doesn't happen, since stone is denser than water.]

5. The ship in the locks springs a leak and sinks. Does the water rise? [No, it falls. Same reasoning as in #4, the ship now rests on the bottom; its water-filled steel frame displaces only a small quantity of water, which reduces its pressure on the bottom a tiny bit more than it would in air, but not so it is weightless, i.e., floating.]

IV. SUBMARINES

A small flask, weighted so the mouth hangs downward, is filled to the brim and allowed to sink. If we blow air into it with a rubber tubing, at first it rises base-up from the bottom, but remains at that depth. Finally, after a certain quantity of air has been blown in, it begins to rise and finally surfaces. Clearly, only when the air bubble has displaced sufficient water to equal the weight of the glass flask unit (Hy11), can it rise. This "blowing the tanks" in a submersible is done with previously bottled compressed air. However, usually, this is possible only down to limited depths. Alternatively, a U-boat can ascend by orienting its depth planes (horizontal rudders) angled slightly up, so they lift due to its forward motion. For extreme depths of over 100 meters (10 atmospheres), the air compressed in the tanks won't get a chance to expand fully and so will displace less water; thus, for the initial lift, we have to use the depth planes.

An experiment linked to this principle is proffered by the Cartesian Diver (Hy12). A small enclosed bubble of air can be compressed by outside water pressure. If the pressure is decreased, the bubble expands, displacing more water, and the 'diver' rises. This gives us the opportunity to relate the exploits of the early daring undersea explorers and of the deep-sea life they discovered.[3]

Experiments in Hydraulics

HY 1 SQUIRTING DISTANCE

a. Two different sized squirters (for example, large plastic syringes, e.g., 20 and 50 cc), with nozzles of the same size (about 3 mm) are taken outside, and squeezed horizontally with equal force by a student; the smallest diameter squirter shoots the farthest.

b. The angle-squirter squirts sideways just as far as it does frontward.

c. A so-called Pascal flask (radial squirter) makes jets of equal size in all directions. [Alternatively, using a glass-blowing torch, melt and blow out little holes in a Florence flask.]

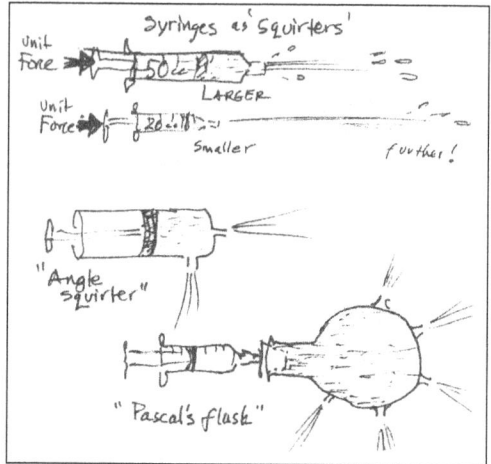

Figure 25. Squirters

HY 2 HYDRAULIC BALANCE

Connect and mount two equally-sized syringes (e.g., 100 ml) with tubing and place weights on each. The weights must be placed exactly on the center of the plungers or they will bind and not give a true result. We also place different (but equal) weights on each; still they don't move. But, if one weight is not a precise match for the other, the plunger under the heavier one will shift down.

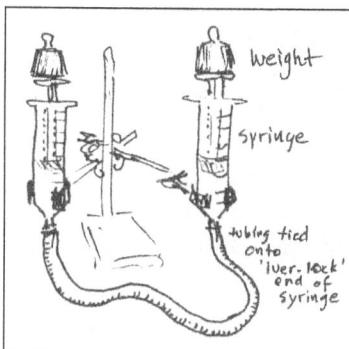

Figure 26. Hydraulic pistons

HY 3 PRODUCTION OF HYDRAULIC FORCES

For one of the larger plungers of Hy2, we substitute a smaller syringe (e.g., 50 or 20 ml). Now we have to use a smaller weight on the smaller syringe to balance the larger. Thus, it is not the volume of the piston & cylinder that matters, but the surface area of the piston (r^2 x _). For an exact measurement, the total balancing weight would have to be corrected for the various individual weights of each piston. For the two 'boxing' plungers (Figure 24), the larger one pushes the smaller on in, since the larger one receives a larger force from the same water pressure, coming from the large jug.

We can also demonstrate a hydraulic jack. Screw in the jack handle (to close the bleeder valve) and pump the lever. Usually the hydraulic oil sits in the reservoir inside the piston casing. Pressure is made in it above the small pressurizing piston, which runs by a small channel beneath the larger piston; this one moves less far, but with greater force—thus creating the useful lift for a heavy car, engine, safe, etc. By loosening the screw of the bleeder valve, the pressure is relieved, and the piston can slide back in.

HY 4 DEPTH-PRESSURE

A pressure-measuring apparatus specially designed for these experiments consists of a piston at the end of a tube. On the inside of the tube we have air; outside, water. The water presses in, and a balancing spring, connected to the piston, holds the piston back against the water pressure. In an aquarium, we measure the pressure at various depths, 10 cm, 20 cm (with pressure measurer held vertically); also for 10 cm with the device held as diagonally as possible.

Alternatively, try attaching a diaphragm, with a long indicator needle glued on it, onto the end of a thistle tube. This can indicate the pressure by the tiny inward bowing of the elastic diaphragm. Although pedagogically a bit less clear, it may be easier to construct.

HY 5 PRESSURE AT DEPTH (PASCAL'S VASES)

Pascal's vases[4] are mounted and gradually filled; a special measuring apparatus (spring force indicator) is used to show the force on the bottom end cap when it comes off and water runs out. Since the bottom openings of all cylinders and funnels have equal areas, the force on the end cap corresponds exactly to the pressure on the cap at the point it leaks. We see that the pressure at depth depends only on the depth of the water column, not on the quantity (volume) of water on top of the end cap.

HY 6 PASCAL'S VASES

a. One very traditional demonstration consists of a series of small glass tubes of various shapes, joined together by a common connecting tube at their bases.

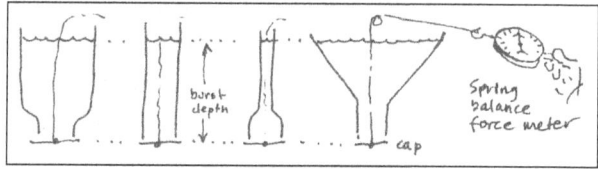

Pascal's Vases

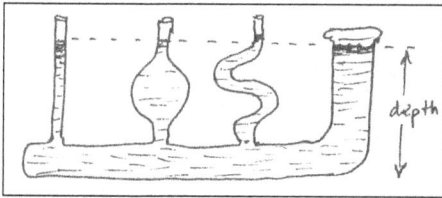

Figure 27. Communicating tubes

Alternatively, we could take various pieces of glass tubing, bend and blow them out over a Fisher burner, and then connect them with rubber stoppers and tubing, perhaps separated by a distance across the classroom. That way we get to see the sloshing of the water inside as it seeks its own level.

b. For another glass tube, we make a bottom cap out of wood, which can be held in place with twine (like the Pascal vase end caps). We weigh the wooden end cap so it will sink when it comes off. Immersed in water, the bottom cap is pressed on by the surrounding water pressure. Gradually fill the tube. When the water level is equal to that outside, then the water pressure which presses out on the bottom cap is equal to that pressing in, and the bottom cap peels off.

HY 7 EXIT STREAMS (MARIOTTE'S CYLINDER)

Make three holes (about 2–5 mm diameter) at different heights in a large (10 liter) cylindrical tank or jug. Mount small (2 cm long, 2 mm dia.) pieces of plastic straws as spigots in each hole and fix them in place with tape or silicone sealant. Filled with water, each spigot will squirt to a different distance, corresponding to its depth. [A 2ft length of 4" PVC drain pipe drilled with 3 or more holes also works quite well.]

HY 8 BUOYANCY

Saw two blocks out of the same dense wood (elm, oak, etc.) so they will have exactly the same volume and weight, something like the following (Figure 28). One after the other, we place each of the well-varnished wood blocks into our water tank and load them with greater and greater loads, until each finally sinks. Record

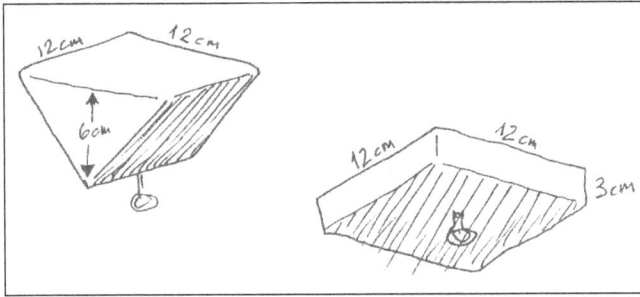

Figure 28. Equal size flotation blocks

the maximum load each will support. We see that each provides the same carrying capacity. Then we measure the dimensions of each and see they have the same volume (and thereby displace the same volume of water when they are just awash and about to sink).

HY 9 ARCHIMEDES' PRINCIPLE

Hook an aluminum block (about 5 cm on each side) or ball to a 500-gram spring balance and suspend it just above a large (400 ml low form) beaker filled about half full with water. The beaker stands on a 500 gram pan balance. After immersing the aluminum block or ball, the pan balance shows a larger weight and the spring balance a smaller weight.

[Note: For hot-shot students, we can pose a challenge: What does this say for weighing in air (or a fluid) as opposed to weighing in a vacuum? The buoyancy decrease in weight, due to air displaced by the object being weighed must be corrected for to obtain a very precise weight—although the correction is quite minuscule.]

HY 10 DISPLACEMENT

Place a previously weighed wood stick into a graduated cylinder. The rise in water level is read off the markings on the side of the cylinder. [We assume that 1 cm^3 water = 1 gram; which is exactly true at 4°Celsius, which was an early basis of the gram]. The floating wood stick is found to displace exactly as much weight of water as its dry weight.

[Perhaps we could retell the story of Archimedes' "aha" experience as he discovered the key to solving the king's puzzle: how to tell if a crown was solid gold or just a base metal with gold plate. Since gold is denser than all other metals common in his day, the problem is one of finding the density (weight/volume) of a very complex-shaped object. Weighing the crown is easy; finding its volume without melting it down to a block is difficult. Archimedes realized that by using the technique of water displacement, he could easily and accurately measure the crown's volume simply by measuring the amount of water displaced.]

Metal Densities	
Metal	Density
Pure Gold	19.3 gm/cm3
Silver	10.5
Lead	11.5
Brass (copper & tin)	11.4
Iron-Steel	7,3-7.8

HY 11 U-BOAT (BOTTLE SUBMARINE)

Set up a bottle (about 200 ml) with a 2-hole stopper and 'keel weights' of coins as follows:

We experiment with how much weight is needed so that the bottle will slowly sink if left free. Using a long and very thin rubber hose, we blow air into it: It will begin to rise, faster and faster toward the surface, finally trailing a stream of bubbles from the second stopper

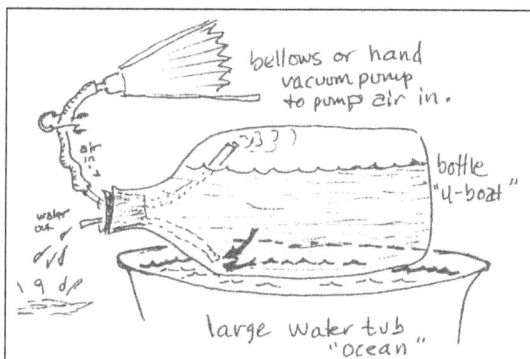

Figure 29a. Bottle submarine

hole. It will remain floating as long as we clamp off the rubber hose. Opening the 'sea cocks' (the tubing clamp) allows air to escape as water rushes in the lower hole, and the bottle submarine sinks to the bottom of the "ocean."

HY 12 CARTESIAN DIVER

The well-known demonstration of the Cartesian diver makes an impressive conclusion. Set up a large, clear jug (10 liters) with a syringe attached either to a side spigot or to a tight-fitting cork at the top. Pull the syringe half way out, to fill it with water. Expel the air by pumping the syringe in and out repeatedly. Take a small (8 mm x 15 mm) test tube, fill it only 80% with water, and invert it in a water trough. Test this "diver" and add air bubbles or fill it more fully with water until it just barely floats. Now fill the jug completely full, insert the prepared diver without losing any water or adding any air to it, and carefully close the top of the jug.

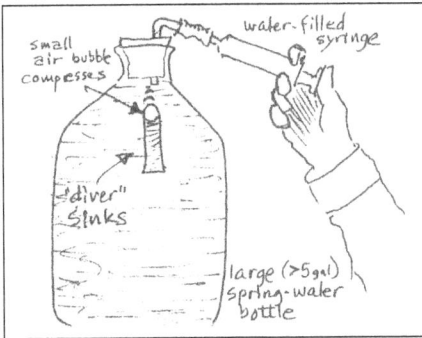

Figure 29b: Cartesian diver

When pressure is applied to the water-filled jug using the attached syringe, the only thing which can compress is the air bubble in the "diver." Compressed, it displaces less water, decreasing its buoyancy as we have learned and it sinks. When the pressure inside the jug is reduced by pulling on the syringe plunger, the "diver" rises dramatically. It is no small feat to just maintain the diver at a given level. [Note: as the diver sinks, the increased pressure will augment the compression of the buoyancy bubble, making it sink faster. Conversely, at depth, a greater reduction in pressure is needed to cause the bubble to expand, causing the diver to finally begin to rise. However, once it rises, the decreased pressure with decreased depth will overshoot the intended rate of ascent, making the diver quite difficult to control at first, until one has gotten the knack of it.]

Grade 8 Aeromechanics

INTRODUCTION

Unfortunately, this topic (like hydrostatics) is often also begun by starting with weight, instead of with pressure. Teachers often focus on the weight of the earth's atmospheric sheath. But, how can we do this in a concrete way, since the density of air decreases gradually with altitude and the atmosphere has no clear upper boundary? A study of this topic solely from the point of view of weight relationships warps the connections even more than for hydraulics; for, in the final analysis, it builds upon the [theoretical] comparison of an air-filled earth with an earth devoid of air. These things cannot be so casually presented if we are to keep in touch with reality. This is not to say that we wouldn't pose such a thought as a possibility to be considered; but, it does not offer *an elemental experience of the phenomena of air*, nor build a reasonable entry point to its study.

I. THE ATMOSPHERE

Already in antiquity, it was noticed that when an air-filled flask was sealed in the lowlands and then opened on a high mountain, **the air rushed out**. And inversely, if recapped at high altitude, brought down to low elevations and reopened, **the air rushed in**. If opened under water, it would even become partially filled. Also, people knew about altitude sickness; and also that cooking done in high, mountainous places would not progress as quickly as at lower elevations. We must take these phenomena to mean that the air up there is *not weaker, just thinner* (more rarefied). Just as in water, the pressure decrease strongly corresponds to the decrease in depth from below upward, so also it appears that in the air there is a **delicate pressure** decrease going upward and the atmosphere becomes **thinner**.

Could there be a space, somewhere in the world where there is no air, only emptiness? Could total nothingness exist on the earth, so there would be no breeze or wind? At first, all philosophical principles and even our experiments seem to say that this is not possible: Nature prevents it. In the Middle Ages, people spoke of a *horror vacuii*—Nature abhors a vacuum. Where there is nothing, God cannot be—that was the problem!

People had whole journals of experiments which were supposed to prove this; for example, pulling liquid up a drinking straw (Ae1). To a naïve understanding—also for the students—this simply has to do with suction. But how is this 'sucking' *transmitted* from the retreating back wall (of the space creating the sucking, i.e., the tongue) to the water's surface? What sort of *pulling forces* pass from the skin of the mouth through the air over into the water?

To a (Renaissance) mechanical way of thinking, it is as though there are invisible strings doing the 'pulling' even though the air has no form. Plainly, between the air and the water there can be no vacuum if this sucking is to operate. People imagined "that water and air, *because of their kindred nature*, make a boundary between each other of such nature that it belongs to both, and they both remain stuck to it as if with glue, so that when we suck out the air in a straw, then the water must follow." (Dijksterhuis) Galileo also debated this point. His Mr. Salvati[1] argued how "the force of the vacuum is distinct from other forces," and then told how it can be measured (Ae1). Water is adapted to this, that "there is no other force of resistance against separation of the parts than the vacuum."

The thought experiment presented reminds another of Galileo's characters, Mr. Sagredo, of an experience with a lift pump over a well. Any pump will fail to work as soon as the water has dropped beyond a certain depth. In response to the questions, the Master in Galileo's book answers "that neither by means of pumps nor by means of any sort of mechanical devices which raise the water by attraction, is it possible to cause this same water to rise a hair's breadth more than 18 Ells (32 feet), and whether the pump be broad or narrow, this remains for these devices the uttermost interval."[2] The members of the discussion find themselves of the opinion that at this critical height, the water column tears like a stick, like a wooden or iron rod which finally ruptures under the force of weight. Galileo had them say only things hold together "like wood:" He called it only a comparison. He saw the actual source of this cohesive force in the mysterious "force of the vacuum."

Historically, we can follow how investigators wrestled in order to understand this sucking—which we naïvely understand as a force acting from a distance—how they tried to understand it as a force passing from point to point. The puzzle is: Why doesn't the pump pull up water higher than 32 feet? Why should Nature abhor a vacuum and hold things together only up to 32 feet? The solution comes only when we think in terms of *pressure* rather than a pulling force. The vacuum does not pull up or draw the water; rather the incompletely evacuated space [with its remnant of air] creates an imbalance with the normal air pressure outside, which then can *press up from beneath.*

Now, back to everyday phenomena, which show the vacuum at work. We see, e.g., that water in a flask (Ae2) will flow out only when air flows in. We are all familiar with small juice cans which pour only when we punch *two* holes in the top. If we don't provide for the air to come in, the water in a jug will only pour out with "glugs" and splashes; periodically it has to "pull in" air (after a momentary expansion of that already inside as the water pours out). In addition to their spigot, storage cannisters or jugs, therefore, have a second small hole to let air in. With the straw, the air hole is built in (the upper opening). An automatic watering jar for animal cages is based on this principle.

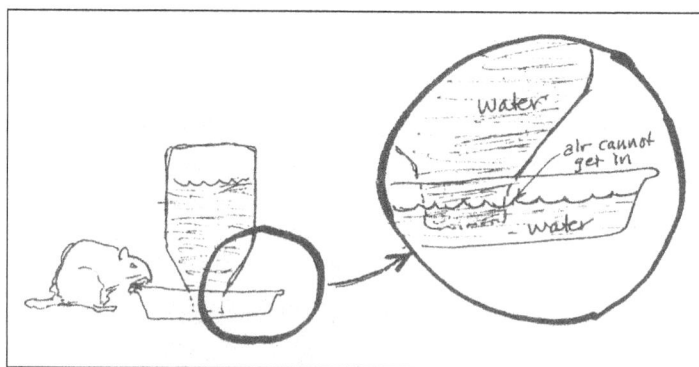

Figure 30. Automatic animal watering jar

Only when the water level in the dish falls below the mouth of the jar will air be able to flow in and thus water to flow out. This outflow refills the dish and stops the flow again. Thus, the water level in the drinking dish remains shallow, even though the whole device has a large capacity.

With these considerations—certainly only in an introductory way at first— we have set before our eyes the actual way air acts on the earth, and as something which fills everything and spreads out in all directions. This elastic, space-filling expansion is clearly the gesture of air pressure, which must be experimentally disclosed first as *pressure,* and only later do we have to do with the *weight* of air. The action of air pressure (not of weight) is plainly a leading thought for the 8th grade syllabus. "For, in a way you will conclude the physics lessons through aeromechanics, through a mechanics of the air, in which everything connected with climatology, the barometer, and meteorology will also come into discussion."[3]

It was already known in the Middle Ages (see above) that a lift pump—like those in use today, driven by hand—would not lift water higher than 10 meters (32 feet). In order to prove this, Berti, a contemporary of Galileo, attempted to make the water rise up long, vertical tubes. His findings are repeated in our water column experiment (Ae3). Since water is never entirely free of dissolved air, we never get a complete vacuum; a tiny residual pressure remains, which pushes the water column down slightly, so, we only achieve a rough value of 10 meters. More precisely, if we include the residual water vapor pressure, the best vacuum we will obtain is a 10.3 meter water column; for higher locations with an elevation of 100 meters (300 feet), however, it will be only 9.9 meters.

In explaining our experimental results, we could say that the "horror vacuii" has a limit: It will support a column of only 10 meters. The width (diameter) of this water column and also its form play no role (see Ae2); it depends only on the vertical distance. We can imagine that we could elongate the little water surface in contact with air formed at the hole in a cork (see Ae2b), extending it to a 10-meter column. The pressure of 1 kg/cm^2 corresponding to the bottom of this 10-meter water column is just balanced and supported by the pressure of the air. So this is the magnitude of air pressure at sea level—precisely the force able to support a 10.3 m column of water (or, more accurately,[4] 1.033 kg/cm^2). Normal air pressure will not support a larger column of water: It will flow down until only 10 meters remains, forming a nearly complete vacuum at the upper end of a closed tube.

Barometers originally measured air pressure in terms of the height of such fluid columns. With the water column, we already have a barometer in front of us. The sea level water column of 10.3 m can fall to about 9.5 meters for extremely low-pressure weather conditions, or else climb to 10.8 m with good, high-pressure weather. The modern units are millibars (mb) or kilo-Pascals (kPa). The average sea level pressure is 1013 mb (101.3 kPa), the highest fair-weather pressure 1070 mb (107 kPa), and the lowest is 930 mb (93 kPa). For our usual rainy-stormy, low pressure conditions, it can only fall to about 100 kPa; and for a continental high pressure weather pattern, it rises to 103 kPa. There is a daily variation around these levels; about 10 a.m. and 10 p.m. the air pressure peaks about 0.1–0.2 kPa (0.7 mm Hg = 1 cm water) higher than in between these times. This rhythm is a reflection of the daily

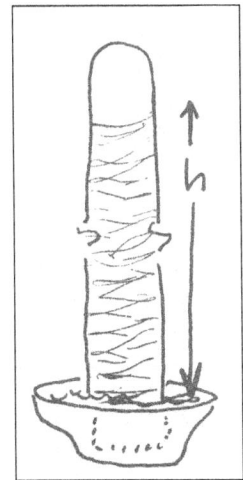

Figure 31.
Water column

rotation of the earth, and how the "sea of air" bulges outward toward the sun, analogous to how the oceans bulge due to the sun and moon causing a tidal rise and fall in the ocean which circles the earth daily.

The mercury barometer functions in the same manner. Only, as we go down in a fluid column in mercury, the (hydrostatic) pressure increases 13.6 times faster than for water; the density of mercury is 13.594 g/cm3. Thus, 'standard' air pressure of 101.3 kPa can support only a 760 mm column of mercury. Also, the 760 mm (30") long mercury tube is a much more convenient length than the 10.3 m (34 foot) long water column. This type of barometer was invented by Evangelistica Torricelli (a student of Galileo's) and was in use for centuries. Even today, we often hear the barometric pressure expressed in mm (inches) of mercury—or else in the alternate unit of Torr. However, a mercury barometer is fragile and troublesome to use.

A contemporary alternative is the aneroid barometer (from the Greek: *aneros* = dry; *barys* = heavy).[5] It is constructed with a slightly evacuated, thin-walled disc-shaped metal can, with very thin, corrugated walls; these walls bend in or out as the pressure rises or falls. A rod soldered to the center of one side is attached through a gear and lever arrangement to a pointer, which rotates so it shows tiny changes in the size of the can. It is useful to have one of these demonstration barometers on hand to show how the pressure changes inside a bell jar. Use an aneroid barometer inside the jar, and briefly lower the pressure by sucking on a tube attached to the jar, causing the mercury barometer (running up through the top) to fall (Ae4). Students can make their own aneroid-type barometers at home out of a bottle with a rubber cap and a straw (Ae5).

Using the various pieces of apparatus described for measuring the pressure of air, we can get a clear understanding that, with them, we grasp only a tiny aspect of reality from the realm of air, since the fundamental phenomenon of the expansiveness of air cannot be perceived. In order to measure air, we have to make something with air absent: a vacuum space, e.g., the space above the mercury column in a Torricelli tube. Since vacuum space is hermetically sealed off from the normal world, it can interact with it only mechanically (or hydraulically through mercury). In this way, we obtain the height of a fluid column, i.e., a length measurement, which can be extrapolated to a concept of the force of the atmosphere. Who would have thought that to investigate the air-filled realms about the earth, we would renounce meeting this selfsame air and deal instead with its absence?! Who would have thought that we would use a 80 cm *mercury-filled tube*, dipped into a dish of mercury and then by measuring its length get to know something about *air*! From the external aspects of these phenomena we clearly derive mechanical ideas. Thus, only the mechanical aspect of air is understood. But astonishingly, this one aspect of the thing—the mechanical characteristics of the air, the pressure—*is exactly the aspect which has a significant connection with weather*. So, this topic is just the place—empirical meteorological rules, based on the barometer—to weave in the mechanical aspect.

In civil airplanes the pressure is allowed to drop to the equivalent of 2000 meters (6500') elevation; it is held at this level at higher altitudes. Then, a positive over-pressure exists inside the airplane. If a window breaks, the passengers will be sucked out by the lower pressure outside. In military aircraft, the cabin pressure is allowed to fall to an equivalent of 12,000 m (39,000') and oxygen is required for breathing. The air pressure is about:

sea level	760 mm Hg =	1013 milli bar (mb)
Mt. Blanc (15,771')	416 mm " =	555 mb
Mt. Everest (8850 m) (29,035')	235 mm " =	313 mb (3 m water)

On such high mountains, water already boils at 70–84°C (Mt. Everest); for every 1500 meter increase in elevation, the boiling point falls by about 5°C from the sea level value. Also, at 5500 m there is only half the pressure, and at about 11 km only about one quarter of the pressure found at sea level. The actual pressures measured on top of these high mountain peaks are a bit lower than those often predicted in tables, which often miss making a correction to account for the fact that the temperature is 10–20°C lower up there and the air is dryer: both require a correction due to the air being denser than at lower altitudes. (Insofar as we consider gravitational acceleration at latitudes other than 45°, yet another correction is needed, albeit a small one). The following table takes into account the effect of the drop in mean temperature aloft. (It is based on the DIN temperature equations for standard atmosphere):

The **Magdeburg hemispheres** are capable of developing a considerable force. This force arises naturally as a consequence of the action of the vacuum pump, developed by Baron von Guericke [*Ger*-ruh-kuh]. Two people have to pull on

ELEVATION-PRESSURE TABLE		
Altitude	Mean Pressure	Water barometer
0. m sea level	1013. mb	10.3 m water
100.	1001.	10.17
200.	989	10.06
300	978	9.94
400	966	9.82
600	943	9.59
800	921	9.36
1000	899	9.14
2000	795	8.08
4000	616	6.27
8000	356	3.62
10000	264	2.69

each end in order to overcome the low pressure inside. With his pump and an altered fire hose nozzle from the city of Magdeburg, Guericke conducted a number of celebrated experiments. In order to understand the dreadful character of this emptied-out space inside the hemispheres, we imagine how it would be if we tried to breathe with our nose and mouth closed or tried to pull up on a collapsed bellows. We introduce a downsized edition of Herr von Guericke's Magdeburg hemispheres, which originally were 52 cm diameter (Ae6). Since a hand pump can act effectively down to about 50 mb, we achieve a clamping force of almost 1 kg/cm^2; for our 10 cm diameter hemispheres, thus approaching a force equivalent to 78 kg (172 lbs)! Since we're focusing on teaching about air pressure and vacuum phenomena and not force, clearly, the force should be given in kilo-pounds or Newton, rather than mass units of kg (1 kp ≈ 1 kg @ 1 gravity). If we open the stopcock a bit, it hisses loudly for a few seconds and the force is gone: Can so tiny a quantity of air possess so large a force? This seems like magic!

Guericke already saw this as air pressure at work [on the unbalanced near-vacuum inside], pressing the hemispheres together. If we use a cylindrical can instead of the hemispheres, then nothing is different regarding the force, though perhaps the volume of air to be pumped out is larger. Mostly, it depends on the size of the sealing flange. Otherwise, if the cylindrical half took up more force than the hemispherical half, such an asymmetric pumped out body would start to move of its own accord, since it would have, e.g., more force on the left than the right.

The considerable force of air pressure can also be demonstrated if we obtain a metal can and pump it out: Usually it will cave in and collapse dramatically (Ae7). We can calculate the atmosphere's force on a highly evacuated telescope tube: 1 kg/cm^2, comes to a metric ton on the extended surface. Models and a drawing of the principles of operation of a traditional farm water pump (lift pump) shows an application of air pressure (Ae8). Guericke considered that the force clamping the hemispheres together was equal to the weight of an air column which extended to the top of the atmosphere, with a diameter the same as the sealing flange of the hemispheres. The weight of the entire atmosphere on the whole earth's surface is proportional to the force on the hemispheres, as the surface area of the entire earth is proportional to the cross-sectional surface area of the hemispheres. Guericke developed this thought primarily in order to be able to think of the atmosphere as having a definite boundary, since he wanted to imagine empty space in between the celestial bodies of the Copernican planetary model. Students in grade 8 should only be gently reminded of these concepts of weight—as already indicated; **more elementary and eloquent is the all-filling air pressure**. We don't describe it as the pressure of the weight of a column of air—in contrast to conventional grade-school textbooks. Rather, we always speak of an all-filling, expanding, pressure-transmitting gas.

The experiment with the two hemispheres was first performed by von Guericke using two teams of horses. In 1663 he performed it in front of a whole circle of onlookers, including Friedrich the Great of Brandenburg. In the Reichstag of Regensburg (1654) Guericke publicly demonstrated in the presence of Kaiser Ferdinand III how the low pressure stored previously in his receiver flasks would pass into cylinders mounted there in the town. It seems one could "bottle nothing"!!

The popular school experiments done in a vacuum under a bell jar (the 'silent' electric bell, bursting air balloon, soap foam, dying candles, etc.) could be deferred until grade 9, in order to make possible a really rich meeting with air pressure and weather phenomena in grade 8. Thus, here, we consider vacuums in opaque, dark containers, more from the outside, from the point of view of the atmosphere. Only in the following class do we really enter into this evacuated space with all sorts of experiments in a vacuum under glass.

Experiments in Aeromechanics

AE 1 SUCTION

a. After sliding the plunger of a 50 cc syringe all the way in, we experience as we pull the plunger out (perhaps with our hand over the nozzle) how great a force arises as nature "avoids a vacuum."

b. Using the syringe, we suction a whole basin of water through a glass tube. Note how the water level in the tube rises effortlessly upward. [Note: Do not be too forceful in sucking up the water, or the incredibly low pressure will cause the water to 'boil' at room temperature. This is not the point, initially.]

AE 2 FLOATING WATER

a. Fill an ordinary bottle with water and invert it. Water runs out with glugs and splashes.

b. Now, insert a single-hole stopper. The water no longer runs out, although a small bead of water in the stopper hole makes a free (unsupported) water surface with the air, the width of the drop: water floating over air!

c. Fill the bottle again, invert it, and slide a glass tube in from underneath. The glug-like pouring ceases and becomes a steady rush of water—as long as the mouth of the glass tube is dry and open.

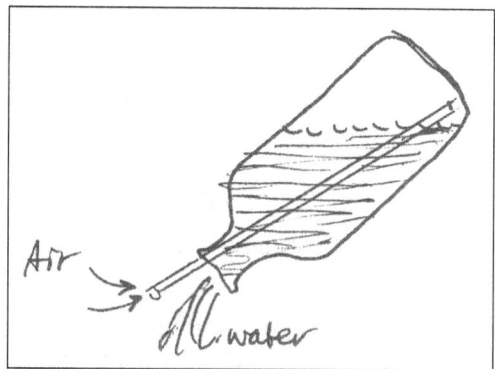

Figure 32.
Water pouring out while air comes in

AE 3 WATER COLUMNS (GALILEO'S PUZZLE)

Take a transparent plastic hose of about 12 mm dia. and 15 m (50') length; immerse and fill it with blue-colored water, which was previously heated and cooled to drive out the air. Ink is not a very good dye, so we use either watercolor or non-toxic pigment from an industrial source. Seal the end with a rubber stopper. Haul the hose up [at least 12 m (35') high] to the second story of the school building using a pulley and a rope.

Now, immerse the lower end in the water basin and remove the lower plug. Immediately, the water column falls to a vertical height of 10 meters above the water level in the basin (as discussed earlier). All along the hose length we

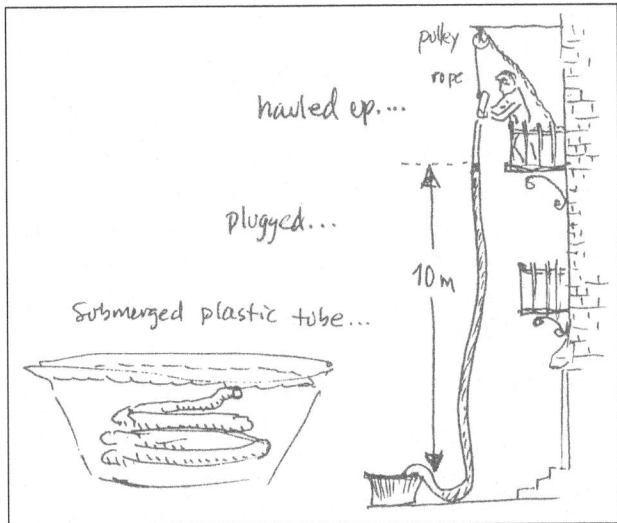

Figure 33. Outdoor repetition of Galileo's water column experiment

plainly see a suction action in that the hose is slightly collapsed, especially in the space just above the water. No matter how we raise or lower the hose, or even hold it slanting, the water column maintains a 10 m vertical height.

Then, with the lower end underwater in the basin below, insert a one-hole stopper and raise the lower end of the hose. None of the water runs out, and the water drop at the stopper hole remains suspended in air. Now, raise the whole hose from above, so the loop on the ground is pulled up into the air. A little bit of the water column runs out. We can mark the height of the upper water surface with a string. Pull out the top cork and see how the water immediately runs out. Finally, we only have to measure the length of the final water column, as marked on the hose from the upper knot to the lower end: 10 meters!

[Note: we can now point out how no greater vacuum could exist at the top of the sealed hose; no amount of sucking could cause the water column to rise any higher! The 10 m length is due to the air pressure, now unbalanced from inside the tube, pushing *up from the bottom opening* of the hose. – Transl.]

AE 4 BAROMETERS

[Extracted: Show how its column is 13.6 times shorter than a water column; show how its height is maintained at about 75 cm (29.5") no matter how the column is oriented—just like the taller water column. Keep a record of the exact barometer height over a period of days, and correlate with the weather conditions.

If possible, show an aneroid barometer; note the construction of the thin metal bellows and the lever action used to magnify the tiny in and out motion of the bellows to give a large movement of the pointer. If it is a recording barometer, allow it to record for several days while keeping a weather journal. Then compare the daily weather patterns with the pressure record. Note how the dense, cool air of fair weather gives a higher pressure, while the less dense, warmer air of summer thunderstorms gives a lower reading. – Transl.]

AE 5 HOME-MADE BAROMETER

Popular hobby/science activity workbooks often describe many delightful experiments. Unfortunately, they often also give explanations laced with a mechanical theory of gas pressure in which hypothetical particles play a central role—at best an unpedagogical mixture of phenomena and theory which obscures an unbiased view of the world. However, this doesn't diminish the utility of the experiments, such as the following:

With strong string tie a 2" x 2" part of a torn balloon over the mouth of the bottle. Glue a thin wooden stick onto the edge of the membrane made thereby. Place the barometer in a place with a steady temperature and mount a scale positioned at the end of the indicator stick. With the daily variations in pressure, the stick shifts up and down: If the pressure increases with good weather, the rubber membrane is pressed in and

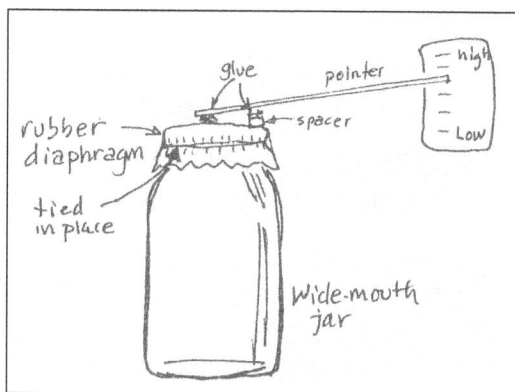

Figure 34.
Home-made milk bottle barometer

the indicator end of the indicator rotates up; if the air pressure falls, it presses less forcefully on the membrane, allowing the air inside to bulge it out and the indicator falls.

AE 6 MAGDEBURG HEMISPHERES

a. Demonstrate especially the effort required to pump out the hemispheres with a hand pump, as well as the incomplete evacuation achieved by sucking with our mouth on the stopcock. Even a rather weak student can eliminate the force holding the hemispheres together with their little finger by simply turning the handle of the stopcock (allowing air to rush in). Note the loudness of air rushing in and the duration as a measure of the degree of evacuation.

b. Hang up a (clear) plastic hose over a hook near the ceiling, so one end dangles in a water bucket on the floor. Attempt to raise a water column by sucking by mouth. We can achieve a maximum of only about 2 meters (6.5') or about 1/5 atmosphere, hardly a vacuum!

AE 7 CRUSHING CAN

If possible, locate a metal can with flat sides (like a turpentine can); plastic jugs can also be used, although they will often split, ending the experiment. Pump it out using a hand vacuum pump. This demonstration is even more impressive if we use von Guericke's method: Connect the can to a large, heavy-duty (vacuum-proof) steel tank (the "receiver"), previously highly evacuated. As soon as we open the stopcock in the connecting vacuum tubing, the can dramatically implodes. [Here we are trying to get yet another experience of the tremendous power of unbalanced natural atmospheric forces. In grade 9 we will study how people cleverly applied and harnessed these atmospheric forces in steam engines, etc.]

AE 8 WATER PUMP

A small glass model of this type of lift pump, usually having a piston sealed with a leather or rubber gasket, is easily obtained from a science supply firm; they will actually pump water. It is interesting to note the simple arrangement for check valves: Usually they are just a glass bead held over a hole by gravity. It is also of value to note the position of the check valves, the water levels in the inlet tube and over the piston, and the direction of motion at the two stages of the pump cycle. This could result in a series of drawings showing how it operates.

Grade 8 Acoustics

INTRODUCTION

After introducing the creation of tones with musical instruments (in grade 6) and also introducing the intervals and frequency (in 7th grade), 8th graders should now investigate the role of the physical air itself. This is related to another topic, also given for 8th grade physics: aeromechanics. Although it's a bit difficult to really grasp the concept of pressure waves in air and its implications, this topic has numerous related themes. Thus, we defer a portion of it into grade 9, where the overall connections can be worked through, partly as a review of the groundwork laid in grade 8. This also has deeper reasons, since the whole subject of pressure waves is also studied in the 9th grade, in connection with electro-acoustics— especially in connection with the telephone. The goal [in grade 9] is to explain technical apparati and what they can do; a pure consideration of Nature is shifted into the background. Thus, by grade 8 we have already emerged from a simple consideration of natural connections; even in grade 7, we already began to form our explanations in terms of material-causal concepts, e.g., the wooden frequency measuring machine. This is carried further in grade 8. This mode of explanation tends to pass beyond qualitative observation into technology. So, it is appropriate to carry it further into 9th grade, especially since the quantity of related topics cannot be totally encompassed in 8th grade.

I. OSCILLATIONS IN ENCLOSED AIR SPACES (STANDING WAVES)

In contrast to the vibrations of plates and wires, the oscillations of enclosed air are quite complicated. The air not only moves to and fro, but undergoes a transient compression in some places and a decompression in adjacent ones—all in very rapid sequence. It is worthwhile to flesh out these ideas into detailed concepts: How is the air actually stimulated to make these oscillatory movements?

If a person blows into the mouth of a bottle, only if we blow sharply across the neck/opening does the bottle produce a tone and resound. Actually, an **unstable airstream with eddies and vortices** arises there, which fluctuates rapidly in and out of the tube (see Figure 35). The frequency of the tone produced has exactly the same rate as these fluttering eddies. As the vortex momentarily passes into the opening, it produces first a compression wavefront; the air flows down or away from this zone of pressure toward the bottom, which contributes to the growth of the vortex. Then the higher-pressure wavefront moves back up the tube, finally interfering with the inward-turned eddy, causing it to decay and eventually fall outside the mouth:

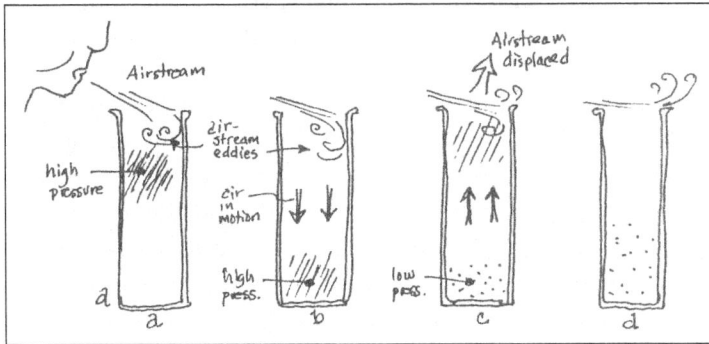

Figure 35. Oscillating air-jet causes a tone in a tube

The size of the bottle corresponds to the size and the speed of vortex-formation and decay. Thus, blowing gently into a large bottle produces a deep tone. Blowing harder, we can also get higher tones (overtones); the vortices are smaller and oscillate more rapidly. (see literature on fluid flow.[1]) The complete details of vortices and standing waves in air-columns are too abstract even for 8th grade students, so we will touch on it only lightly in this block. Instead, **we will investigate in detail the regularity of oscillating air inside a tube**; either considering the ridges of lightweight cork-powder formed inside a glass tube ("Kundt's tube" apparatus: Ac1), provoked by the slow-oscillating metal stimulating rod or else via the water-filled tube stimulated by a tuning fork held over the open end (Ac2), which makes resonant tones at particular lengths. Either of these experiments may suffice; the water-tube is easier to improvise, while the Kundt's tube—though experimentally more challenging—shows the effects much more graphically.

In a water-filled tube, we let the water surface drop down slowly but steadily; for a given length tube, we will hear one tone stand out (resonate) at one (or possible two) water levels. Clearly, the vibrations from the tuning fork have been impressed upon the air inside the tube which now resounds, although only at particular air column lengths. From the length-tone relationships observed, we see that the air apparently requires only a little space for resonating at higher frequencies and a larger space for lower tones. It happens that for deep tones, the compressive wavefronts form in a relatively slower sequence (have their nodes further apart) and traverse a greater distance than with the higher tones; conversely, higher frequency means a rapidly traveled air-path and only a short tube required. Our observations from our investigation will usually confirm this. That it really is only 'light' air which oscillates in the tube can be shown by the fact that the tone ceases quickly when the stimulating tuning fork is removed (just as occurs with a flute, when the musician stops blowing). In contrast strings continue to resound for a while, much longer the thicker and more massive they are, and a very heavy tuning fork will ring for a remarkably long time.

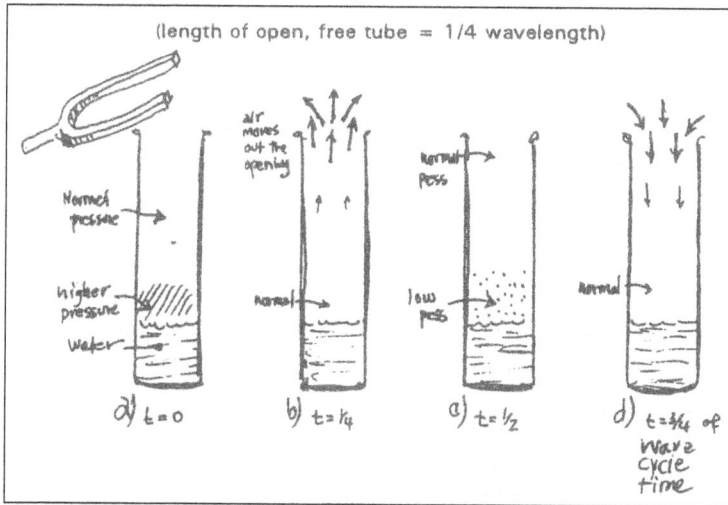

Figure 36.
Movement of compressive wavefronts in enclosed air in a tube

In stage (a) (see Figure 36, above) the high-pressure wavefront is just at the water surface; the water column is relatively heavy and massive, so it acts like a solid reflecting wall.

One quarter osculation later (stage b), the air has now achieved a good velocity; and, although that place is now at normal pressure, just now the air is moving at maximum speed and will overshoot an equilibrium. This occurs partly due to the (tiny) inertia of the moving air itself and, to a smaller degree, because the oscillating tuning fork tine is now moving away from the mouth of the tube. This also works (weakly) to pull the air out of the tube. (In order to carefully work through how the air oscillates, we will not focus so much on the tuning fork, but on the air.) First we consider the fact that at stage (b), normal pressure now exists at the tube-mouth, while the air there now has a maximum velocity. Meanwhile, the maximum change in pressure (decrease to 1 atmosphere) occurred at the water's surface, and the velocity of the air there is least. Another quarter cycle later (stage c) and low pressure now exists at the water's surface and now the air is reversing its motion.

In the next phase (stage d), it follows that there is a maximum of air motion at the open end, and momentarily there is normal pressure everywhere. Naturally, next we will again find the first stage (a) with higher pressure at the water surface, then stage (b), etc.

Wavelength is a new concept which we can develop here, specifically, an explanation involving how an oscillation of the air can be caused or stimulated by the beat of the tuning fork only if the length of the tube is 'tuned' to the tuning fork's pitch. This length corresponds—as we will see—to ¼ wavelength of the tone. If we lower the water level to three parts of the free tube length (air column now ¾), the tube sounds again. The individual stages are as follows:

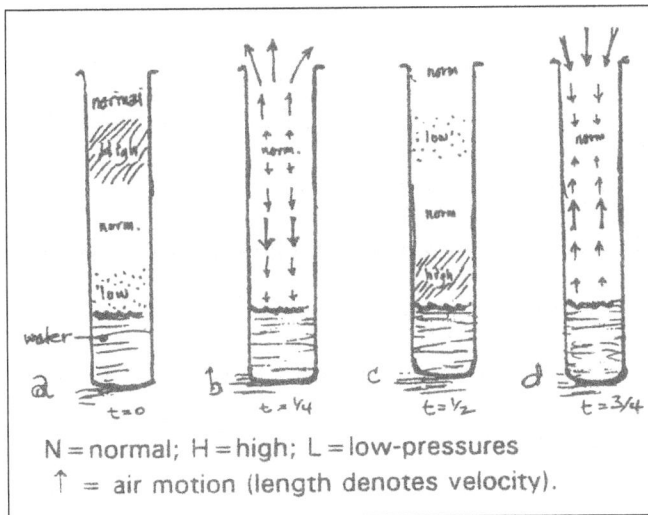

Figure 37. Wavefronts in a long tube

One complete wavelength is the distance from one pressure maximum to the next; from the high-pressure crest to the adjacent low-pressure trough is ½ wavelength. For a closed tube, the experimentally determined tube lengths corresponding to these wave phases are: ¼ and ¾ wavelength.[2] This is similar to the relationship holding for a vibrating metal rod fixed at one end (nodes occur at the fixed or closed ends, and antinodes at the free or open ends) The tube will again resonate at ⁵⁄₄ wavelength (if the tube is long enough). With these demonstrations the students get an **experimentally grounded concept of wavelength** for sound-pressure waves in air. As can be shown with higher or deeper tuning forks, these lengths are related to the frequency of the source tone in a definite way (depending on the density—affected by the temperature of the vibrating air). And, for once, we actually "catch sight of the measured value: it will be equal to about 86 cm or 33⁷/₈" for fundamental mode of A = 440Hz (at 220°C, see below).

The **resonant-tube (Kundt's tube,** Ac1) with its ridges of cork powder (or formerly, in spore-dust from the fungus *Lycopodium*) shows this in a similar way. A maximum of air movement occurs near the inducing rod (or sound-inducing tube) as shown by the powder being shifted away from this region. Minimum motion occurs near the movable plunger, at the other end of the tube. There we see the powder accumulate into a ridge. Such a maximum of motion is called a movement peak (or an *antinode*) and the minimum is called a *node*. At a peak, wavefronts moving from each end meet in phase there (air in each front is moving in the <u>same</u> direction at the instant the fronts meet each other) so the greatest velocity occurs there. At a node, the movement is null but the pressure has a peak there; i.e., an oscillation exists there between high- and low-pressure. (Fronts meeting there are moving in opposite directions at the instant of meeting, so no net motion occurs.)

The *peaks* and *nodes* are certainly contemporary technical terms, although there may be differing opinions regarding their value in such an introduction. In any case, where the movement is at a maximum in Kundt's tube, the powder is chased away; where the movement is at a minimum, we get a piling-up or a ridge. The distance from the end plate to the bare area (motion maximum, peak or antinode) is ½ the wavelength—as in the water-filled tube. The Kundt's tube apparatus can also help make the wave concept more experiential, since it makes it visible to the eye and also resembles the Chladni figures we studied earlier (see grade 6), although the lovely star-like rosettes we saw there are now rigid, measured lengths.

II. WAVES IN FREE AIR (TRAVELING WAVES)

In a large but enclosed space, the oscillations of the air and the nodes and peaks remain stationary; but, in free air, they travel forward. Each wavefront influences the adjoining air; this influence spreads out in expanding spheres of influence, right out to the periphery. We can see something similar (in 2 dimensions) in the familiar way concentric waves spread out after we toss a stone into a still pond. Pressure waves in air exhibit a decrease in intensity when spreading out; the sound gets weaker with distance. However, the decrease occurs faster in air (than water), since the wave-front not only spreads out in a plane (circular waves), but through the whole of space (spherical waves). But what is it that actually travels from the tuning fork to our ear when we hear a tone? What does this wave actually consist of? The moving pattern!

In **water**, waves are formed from the rising and falling of the water's surface. The water rises up making a wave "crest"; when it subsides, we have a wave "trough." In contrast to air, however, this motion of water occurs across (transverse to) the direction of spreading (propagation) of the wave. Since the actual **motion of water** occurs **up-and-down**, *transverse* to the direction of travel, **we call water waves transverse waves**.[3] But, the water *doesn't experience a net movement* in this direction; without wind, waves on the high seas will not shift a boat away from its station. No net movement occurs as long as the water is deeper than about 4 wavelengths. Only when the bottom is shallower than this depth, will breakers occur in shallow water nearer shore. There, the deeper part of the wave motion runs into the shallowing bottom; these breaking waves will drive floating objects toward shore. This is the portion of the waves we normally experience at the beach!

The water wave-height decreases as the circular wavefront grows larger, i.e., as the circumference along which the wave spreads out increases. Also, water wave-crests are generally higher at their center than near their flanks. This is also the case with so-called rope waves, which are seen when we sharply jerk the free end of a suspended rope.

Unlike water waves, air waves cannot be imagined to travel with a transverse (up/down) vibration; there is nothing to move sideways, corresponding to the water's surface. The air in our environment generally exerts its pressure against every boundary surface (although we usually don't notice this air pressure, since it presses equally on all sides of every surface—see Pascal's principle in hydromechanics). In contrast, the water pressure at its upper surface is zero: It doesn't attempt to overflow this boundary. If a portion of the water is raised above this surface, then the pressure underneath this zone is higher than the surrounding. The wave crest has the tendency to flow down and to the sides—the elevated water crest falls. Due to its movement (inertia), it exceeds its goal and goes further to form a trough. Thus, in a basin, an oscillation will arise (standing waves), on open water, this oscillation will affect neighboring water, spread out and the disturbance will travel out as waves. Given such pressure relationships acting across (transverse to) the wave motion, a transverse wave is possible in water. (In a simplified form, we might speak of the heaviness of water.[4])

But, in air, there are no such restorative forces. A portion of air pushed in any given direction will simply blow on further; it will never rush back to its previous location. **A transverse wave is not possible in air.** The expanding air pressure-wave cannot consist of a sideways elastic rebound of air (transverse to its direction of propagation). How then do waves occur in air? Consider our experiment (Ac2) with the water-filled tube. At certain lengths we heard the tone resound, i.e., we heard the air in the tube resonate; a sound wave must have moved out from the tube mouth.

From the discussion above, we know that such a wave-front at the mouth can only vibrate in the same direction as it travels. So, the air must have been alternately compressed and rarefied in this direction. Thus, the air waves in the tubes in Ac1 and Ac2 can only consist of motion to and fro <u>in the direction of travel</u>. When the air moves toward other air masses, higher pressure results. But now, in contrast to the stationary oscillation inside the tube, beyond the mouth of the tube, in open air, the sites of high and low pressure can travel. In any one spot near the pressure source, a constant alternation of high and low pressure occurs. Every site in the wave field undergoes such an oscillation—like a buoy on the ocean, only faster. At any instant within the wave field, we would find a pattern of concentric spherical wave-fronts, consisting of alternating high and low pressure shells—separated by about 40 cm, for concert pitch A. A bit later, all these spherical wave-fronts would have expanded further, in proportion to the velocity of sound (enlarging their diameter by about $2(v)(t)$, where v = velocity of sound and t = time interval). Whether or not we actually work through such computations in class, the teacher should probably work them out himself for a sample case, in order to be ready for questions with concrete examples.

Resonance between two tuning forks can be used to demonstrate the velocity of sound in a clear way (Ac3). The question always arises in this demonstration: "What strikes the second tuning fork, causing it to move?" We can answer: "The air!" The impact of sound waves must not be very strong, since our skin (placed in near the first tuning fork) feels no strong impact from the sound waves produced there. **How then does the air produce a sharp impact like we did using our rubber mallet? By repeated blows!** In each second, it taps the second tuning fork tines gently 440 times, and certainly just at the right instant. This is decisive: Just as a tine of the second tuning fork is moving away from an approaching low-pressure front, it will be gently suctioned back by the low pressure. And, in the reverse situation, when it is moving toward the first fork, an approaching high pressure crest produces a little bit of resistance and pushes it back Since both forks are tuned to the same frequency (usually by adjusting a rider on the tines of one), the weak suction and pushings of the pressure waves happen gently thousands of times per second, and just at the right instant. Thus, the second tuning fork will usually resonate. This can be summarized in a drawing as follows:

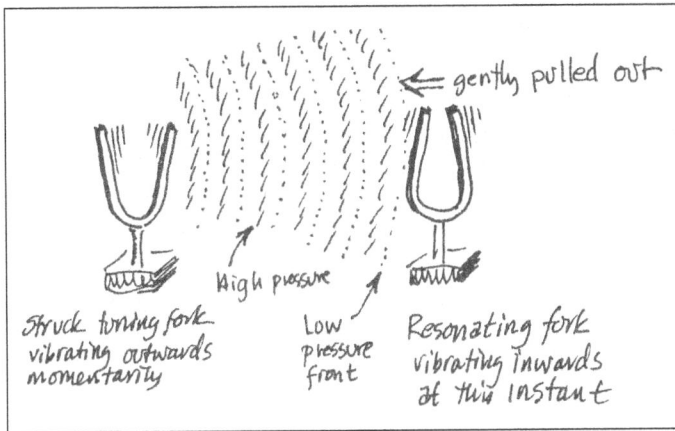

Figure 38. Sympathetic vibration of tuned forks

The perforated siren disk is another good example of the creation of tones via repeated collisions (Ac4). The air stream will be interrupted by the metal plate between the holes. The air moving away on the other side of the disk, due to the previous puff through the preceding hole, will experience a momentary back-suction due to the interruption of the air stream. Just afterward, another hole arrives and the air stream jets through again; this causes a new pressure wave. These successive pressure waves and interruptions create an expanding pattern of alternating wave-fronts in the air, which are heard as a tone with a frequency corresponding to the rate of revolution of the disc and the number of holes passing per second (see the calculations with our wooden frequency-measuring disc in grade 7).

III. VELOCITY OF SOUND WAVES

As a child, everyone has marveled at how it happens that, when we see a sound created far away—for example the splitting of a branch—we usually hear it only a bit later. With this delay, we never think that the distant sound happened when we finally heard it, saying: the event is delayed some seconds, until I experience it, as if, e.g., with thunder we think: At this moment it thunders, now (later) comes a rumble. It is mostly an intellectual superstructure which focuses on measuring the interval between lightning and thunder. So, we never hear it said in fables and fairy tales: "as quick as thunder," but "as quick as lightning."

Actually, such situations involve something quite different than calculated time delay. The shock wave does not just move out unchanged, with some time delay (calculable, based on its velocity). But, it actually **becomes transformed** during travel. The initial sharp crack becomes a dull rumble in the distance. A clear sound not only becomes weaker but sounds completely different in the distance. This inner transformation requires time—*transformation time*, not travel time. The entire surroundings affect it; the sound alters through its passage over the landscape. Nonetheless, inside the classroom, we are constrained to deal with a more or less technically calculable velocity (speed of travel). Echo and reverberation should be studied before focusing on time-of-travel measurements using a meter stick or stopwatch.

As an initial **echo experiment**, we could have the students stand spaced out along a sound path and ask them to wave their arms as soon they hear the sound we make far away with a "banger" (Ac5), in order to show how they wave in sequence. Then we could talk about the characteristics of the echo. (The minimum distance from the wall where we can distinguish an echo is about 30 m or 98 ft). Air waves clearly will bounce off a surface and reflect—similar to the well-known water waves reflecting off a quay or breaker. The entire wave train runs back in a ordered way, so that the sounds stay in sequence and words are still intelligible. The deeper frequencies of human speech sounds in air (about 100–200 Hz for a man) have a large wavelength of 2–3 meters and can still reflect off a boundary/barrier with apertures of up to a meter, e.g., the edge of a forest.[5]

Thus, we hear the deep tones favored in the echo wave-front. In great halls or churches, we hear the so-called resonance of the sounds, particularly under a stone ceiling, and also in long stairways. Each onset (of the echo) is often followed by blurred reverberations, often lasting several seconds; these are reflections off the stone surfaces. This creates poor acoustics. Speech too becomes difficult to understand under these conditions. A 'floating-echo' is heard in circular, low-ceilinged subterranean halls, such as the Mariner's commemorative hall in Labo (Kiel). There, every footfall is heard repeated about seven times, in fading echoes with over 2-second intervals—as if the entire room was clapping in time. So, in concert halls, the walls are deliberately made to be poor reflectors by using sound-absorbing paneling, since it is very disturbing if we hear several delayed echoes after hearing the initial sound of the music;

Measurement of sound velocity is then worked out from observations over a measured travel-path (Ac6). The value found is naturally subject to a moderately large error, mostly in the time measurement. The precise value in air @ 0°C = 331.3 m/sec or 1193 km/h (741 mph). This velocity is not observable or measurable with continuous sounds, nor with music or speech. Rather, these measurements need an abrupt sound, so actually they measure shock-wave velocity.

The velocity (v) increases with temperature (T) according to the formula:

$$v_t = 331.3 + (0.6)\ T\ \ [v\ in\ m/sec,\ T\ in\ °C]$$

This gives us the following temperature-velocity table:

VARIATION OF SOUND VELOCITY WITH TEMPERATURE		
Temp.	Velocity (m/s)	Velocity (mph)
-20 °C	319	714
0	331	741
+20	343	768
+40	355	795

With explosions and similar shock waves arising from explosions, the non-destructive wave reaches velocities as much as three times higher, up to 1000 m/s. Measurements from antiquity with cannon reports over short distances are therefore often inaccurate. (The oldest usable values come from the Academia del Cimento in Florence in 1660.)

Neglecting the temperature influence of air, we can state: The more solid and hard the material is, the greater the sound velocity inside. From swimming we know that we can hear quite well under water, there the velocity of sound is approx. 1500 m/s. Also, as we know from hearing widely separated blast-shocks in mines, the velocity of shock waves in stone is very high, approx. 5000 m/s (almost comparable to the speed of electrical phenomena).

Measurement of wavelength for a known frequency can be performed quite simply (as in Ac1 and Ac2), based on the velocity of the sound waves. For 1 wavelength:

$$(\lambda)\ (f) = v$$
$$or,\ v = \frac{(\lambda)}{(T)}$$

λ = wavelength in meters
f in Hertz (cycles/second)

Let us stop to consider: In one time interval (say, 1 second), f waves (e.g., f high- or low-pressure crests) pass each point in the soundwave-field, advancing 1 wavelength λ; the locale oscillates at the frequency f (compare II-2 Frequency Measurement, Grade 7). From one pressure crest to the next there is a 1-meter space—the definition of the wavelength. Also, in 1 second the wave crests travel f times their wavelength further, covering a distance of f x l.

The fact that sound velocity is mainly influenced by temperature, and thus decreases with cold, means that wind instruments, e.g., organ pipes in an unheated church, play at a lower pitch than otherwise. Sound waves travel more slowly in the organ pipe, since the air is less mobile. The speed of sound is by definition the velocity of propagation (v) of the wave-front. If this is determined by measuring the wavelength of standing waves reflecting back through themselves, the wavelength (λ) must be corrected for the cold temperature, since the resonant wavelength must be an integral multiple of the tube length; moreover, since (λ) (f) = v, and v is now smaller, therefore the frequency must be proportionately lower (or slower), as if cold air cannot move as quickly.

The **direction of sound** is connected with frequency and velocity differences. Interesting phenomena and experiments along these lines are described in various workbooks, e.g., R.W. Pohl, *Introduction to Physics*, recommends taking a finger-thick plastic tube, laying it on a student's neck and putting one end in his left ear. If we tap it near the marked middle, then the sound is experienced as coming from the left, since it reaches the left ear a tiny bit sooner, traveling in the enclosed tube.

IV. VACUUM

Vacuum experiments form a culmination to all our 8th grade studies on the role of the air. What a vacuum actually is, in contrast to our atmosphere, will be studied mainly in under aeromechanics (the previous topic in the 8th grade block). Here, we simply want to show how the sound of a clock becomes weaker as the air is pumped out (Ac7). Everything is separated and cannot make contact. Where there is no atmosphere, no air, eternal stillness reigns! Sound and song prevail [only] in "the open fields of air," i.e., in a middle, living region near the earth's surface. Whether we go up into the rarefied heights of the atmosphere, or instead go down into the deeps, it becomes silent just the same. In the depths of the seas, it generally become silent. Genuine fishes are dumb—they do not make sounds [in contrast to the beautiful songs of whales, which are mammals and thus rise to breathe air]. Deep down in mines, an oppressive stillness weighs down on us. Although sounds are certainly possible there, what is there of Nature is rigid and dead. Also in the highest atmosphere above 100 km, it is silent too; even a plane flying close by is never heard. Beyond the middle zone of our biosphere, Nature becomes ever more rigid and silent—both in the deeps as well as in the heights.

V. EXPERIENCE OF TONE/SOUND

In the 6th grade we progressed from music through intervals to more shrill Chladni-sounds and to the percussive tones. In 7th, we began with the sterile sound of a tuning fork and came to tearing paper. In the 8th grade, we first examined oscillations in tubes. After tuning fork sounds and the whining or moaning song of the siren, we passed on to the 'crack' of apparatus for measuring sound velocity. The fading ringing of a clock in a vacuum brings the investigation of sound to a conclusion. Instead of aiming at gaining total clarity of mechanical concepts, we should always keep in mind that, in the final analysis, our study of sound deals with strengthening our experience of the world of sound; to make it fuller and more ripe by understanding—experientially 'standing within' a wider part of the world.

Experiments in Acoustics

AC 1 WATER TUBES

An alternate enclosed air-column can be set up using a tube which can be filled to various levels with water. We make the water level climb slowly by raising the reservoir flask and opening the clamp slightly.

Now, using a tuning fork which we repeatedly strike to keep vibrating, we determine at what levels (volumes within the tube) we hear an audible increase in the tone ("resonance" in the tube). For the chosen tuning fork, mark these water levels inconspicuously beforehand on the tube wall, so we won't miss them in the experiment.

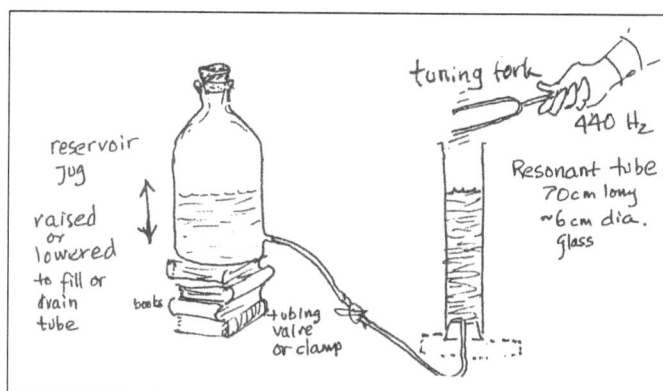

Figure 39. Resonant water tube apparatus

AC 2 POWDER TUBE (KUNDT'S TUBE)

Kundt's tube for making standing waves visible in powder is simply a long glass tube of small diameter, about 3 cm (1-3/16"), fitted out with an oscillating rod (the exciting rod) from a science supply house and arranged as follows:

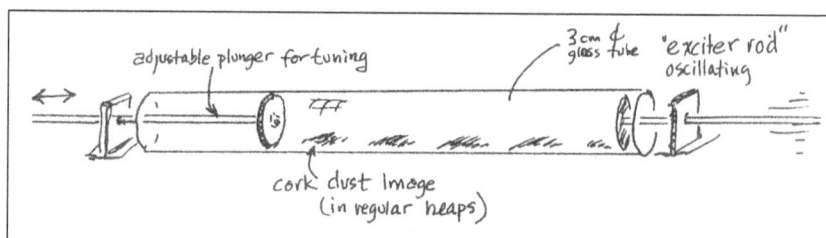

Figure 40. Kundt's tube apparatus

The plunger should be slid out so its distance from the open, excited (tuning fork) end is an odd multiple of ¼ wavelength (at the frequency of tuning fork being used): then the pattern of ridges in the powder will appear as faint ridges. [For example, using a 2 cm-diameter closed tube, a 440 Hz fork should resonate at 18.8 & 56 cm; a 256 Hz fork at 31.4 & 99 cm; a 100 Hz fork at 84.2 & 257.2 cm—first and third harmonic positions.] The position should be determined ahead of time experimentally and marked on the glass tube. The lycopodium powder (very light spores of the lycopodium mushroom) is sprinkled into the length of the tube using a long paper trough, which is tapped gently (with a pencil) to pour out the powder, while rotating the tube a little until the powder begins to slide about; now, it is fluffed up enough. The stimulating rod is bowed with a rosined bow to make it "sing." [Experiment to get a good, high tone]. Alternatively, instead of a stimulating-rod, we could use an aluminum tuning fork of suitable frequency (obtainable from a science supply company) against a funnel covered with a rubber diaphragm.

AC 3 RESONANT TUNING FORKS

We explore the phenomenon of "sympathetic vibration" by placing two similar tuning forks on sounding boxes, setting one in the front of the classroom and position the other nearby with the opening of its box oriented toward the first fork, but some feet away. Now, strike the first fork, allow it to sound for a few seconds, then damp it by grasping its tines: The sound continues! Where does it come from? Finally we see (by also damping the second fork) how it had been sounding in resonance to the first. Now, listen to and then damp this *first* one again: We hear *another* tone damped, showing it was coming from the first fork *again*, which had picked up the tone from the second one (now damped). This trading by resonance can go on for quite a while if we listen carefully.

Alternatively, if we produce a tone with a siren disc driven at a constant speed by a speed motor (or a manual crank), we can even stimulate the tuning fork to resound this way—if the pitch of the siren is correct.

AC 4 SIREN DISK

By blowing air against a rotating metal disc with several appropriately spaced rings of holes, we can produce a whole series of tones and intervals. For a constant rotational speed, the frequency is in proportion to the number of holes in the series at the rim. Thus, this investigation can be a further foundation for the experiment in grade 7 (frequency relationships of the intervals). Instead of an air jet, we can also produce the tones by holding the corner of an index card against the rotating holes—like the toothed edge of the wheel we used in grade 7.

AC 5 EXPANSION OF SHOCK WAVE

Make a "banger" apparatus out of two long boards, hinged together. Paper flags on the ends allow us to tell when the boards snap shut, even at a distance of over 100 meters (328'). Outside, we have four students stand at intervals of 50 m (160'), 100 m (3287'), 150 m (492') and 200 m (656'), who will show by bringing their hands together above their heads when they hear the "crack." (Previously discuss how it arrives a fraction of a second later than the visual cue of the closing flags.)

AC 6 VELOCITY OF SOUND

Have a few students stand 200 m (656') or more away from the banger, equipped with stopwatches (and practiced in their use). Measure the interval from the visible closure to the audible report (should be about 0.56 sec). Then measure the distance by rolling a bicycle wheel along the course and recording the number of revolutions. (The circumference of this wheel has been previously measured by rolling along a tape measure.) Note: in a wind, the experiment must be done in both directions and then averaged.

Note: To find how far away lightning is, you can count the time delay between the flash and the initial high crack. The low-pitched rumble travels more slowly. Then, multiply by 0.2 miles for each second delay (0.35 km/sec). Why? distance = rate x time, and the velocity of sound in warm air = 0.3461 kilometers/hr (770 miles/hr), which we converted to per seconds.

AC 7 VACUUM

While the space inside a vacuum bell jar is pumped out (preferably by pumping out the water using a hand pump), the sound of an electric bell grows slowly weaker. So that its sound is not transmitted through the jar to the sealing base, it is suspended from the top by rubber cords. The vacuum jar (bell jar) undergoes powerful forces which press inward when the air is pumped out (see aeromechanics). Therefore, it must be treated very carefully; it must not be stressed, flawed or have *any scratches* (which would be sites of weakness) in order to successfully withstand the hundreds of kg of unbalanced force when under vacuum.

Endnotes

OPTICS

1   As opposed to **back**-lit, which produces the yellowish-red colors discussed in the preceding paragraph. – Transl.

2   For some chosen viewer-lens distance, but held constant; the lens-object distance is gradually increased.

HEAT

1   Imbalanced forces and mechanical movement made possible through a suitable arrangement or machine parts, is a theme which lies at the center of heat engines, as explored in 9th grade. – Transl.

2   The attempt to deal with phenomena of heat using statistical, mathematical physics was a major and decisive chapter in the 19th century development of modern science.

3   See the flyball governor invented by James Watt to control the steam supply to his heat engine, studied in grade 9.

ELECTRICITY

1   See Grade 11 physics discussion of potential "fields" with Schrodinger's quantum mechanics; the electron is no longer conceived as some sort of particle (tangible thing) but rather as an n-dimensional probability wave field.

2   For a more complete history of the development of the telegraph, refer to Project Gutenberg eBook: Heroes of the Telegraph by John Munro, www.gutenberg.org.

3   W. Harrar, *Seven Years in Tibet*, 1977.

4   Heroditus, Book 8, Chapter 98.

5   Compare McLuhan's global village and its electronic promise of liberation with the alienation mentioned here. Also, contrast the isolated citizens of media net with a communion of cosmopolitan, free individuals, formed out of personal contacts, transcending nationality.

6   Thanks to H. von Baravalle, *Physics as Pure Phenomena, Vol. II*, Bern: Troxler verlag, 1954, p. 207.

7   See Sargent-Welch, Fisher Scientific, or Frey Scientific supply catalogs.

HYDRAULICS

1   Technically, the Newton is a unit of force: kilogram is a unit of inertial mass. One kg exerts a force of 1 Newton only under specified conditions of gravitational acceleration, thus the qualification of altitude and terrestrial position.

2   The force of weight (heaviness) is not the phenomenon, only the acceleration toward the earth,

and (at rest) the supportive force. Naturally, this is yet to be explained later (in 10th grade). Nonetheless, it is possible to think about what we have developed here in terms of the conventional force of weight of solid bodies. But we should carefully avoid also considering fluids in terms of weight since with fluids immersed in fluids, where is the lifting force? Therefore, we have based all this on pressure. Hermann Bauer has written some essential thoughts about weight in *Mathematical-Physical Correspondences*, No. 100, July 1976.

3   See: books on J. Costeau or *The Sun beneath the Sea* by J. Piccard & William Beebe, 1977.

4   See any general science supply catalog under this name.

AEROMECHANICS

1   "Dialogues Concerning Two New Sciences," Galileo Galilei.

2   Ibid.

3   Rudolf Steiner, Second Curriculum Lecture, September 6, 1919

4   Naturally, we mean the kilopound, the weight of 1 kilogram at the surface of the earth, i.e., under a gravitational acceleration of 1 g; it is not necessary, however, for the students to differentiate between inertial mass and weight in this block.

5   The meaning of the word *barometer* (weight of air meter) insinuates a one-sided interpretation in terms of weight, not in terms of the more fundamental pressure. The word *manometer* (pressure meter) is usually used for instruments measuring pressures above atmospheric, e.g., in water pipes. A barometer gives the absolute pressure of the air, a manometer the difference from normal pressure.

ACOUSTICS

1   *Sensitive Chaos*, Theodor Schwenk; *Sourcebook of Experimental Physics*, Bergmann-Schafer, Berlin: DeCruyter Verlag, 1965.

2   Technically, the oscillation (resonant column) extends about $0.5 \pi r$ beyond the opening (r = radius of the tube); therefore the experimentally determined length agrees only within 1 cm of the effective column length, i.e., falls short of the length for that frequency by 1 cm.

3   Technically, the water surface moves in a vertical circle.

4   See Hydromechanics of the Barometer under supplemental topics.

5   Long waves will pass around a corner for the same reason—in gulleys we hear the deep tones more. The very short waves of higher tones are transmitted more linearly, in contrast (ultra-echo).

Selected Bibliography

Duit, Reinders, et al. "Everyday Concepts and Natural-Science Teaching," 1981.

Feyerabend, Paul. *Knowledge for Free Human Beings*. Suhrkamp Publ., 1981, p. 18.

Fischer-Wasels, Horst. "Non vitae, sed scholae discimus?" in: *Die Hohre Schule [The Upper School]*, 9/78, p. 339.

Fromm, Erich. *To Have or To Be*. Basic Books, 1976.

Heisenberg, Werner. "Physics and Philosophy," 1959.

Jung, Walter. "Phenomena, Concepts, Theories: 3 Theses of Scientific Theory and Didactics in Physics," in *Physics Teaching*, Vol. 16, May 1982.

Jung, Walter, see Duit, et al., 1981.

Kinzel, Hebnut. "Scientific Knowledge and Human Experience," in *Biologie in Unserer Zeit [Modern Biology]*, Vol. 9 No. 4, p. 112 (1979).

Schumacher, E.F. "A Guide for the Perplexed," Harper and Row, 1977, p. 166.

Silkenbeumer, Rainer. "Model Schools – School Models," Hannover, 1981.

Steiner, Rudolf. *Philosophy of Freedom*. 1918, 1984.

Weizsacker, Carl F. *The Unity of Nature*. Hanser Publ., 1971, p. 244.

Wieland, Wolfgang. "Possibilities and Boundaries in Scientific Theory" in *Angewandte Chemie [Unified Chemistry]*, Vol. 93, 1981, p. 633.

www.ingramcontent.com/pod-product-compliance
Lightning Source LLC
Chambersburg PA
CBHW051222200326
41519CB00025B/7214

*9 7 8 1 9 4 3 5 8 2 2 4 2 *